SpringerBriefs in Physics

SpringerBriefs in Physics are a series of slim high-quality publications encompassing the entire spectrum of physics. Manuscripts for SpringerBriefs in Physics will be evaluated by Springer and by members of the Editorial Board. Proposals and other communication should be sent to your Publishing Editors at Springer.

Featuring compact volumes of 50 to 125 pages (approximately 20,000–45,000 words), Briefs are shorter than a conventional book but longer than a journal article. Thus Briefs serve as timely, concise tools for students, researchers, and professionals.

Typical texts for publication might include:

- A snapshot review of the current state of a hot or emerging field
- A concise introduction to core concepts that students must understand in order to make independent contributions
- An extended research report giving more details and discussion than is possible in a conventional journal article
- A manual describing underlying principles and best practices for an experimental technique
- An essay exploring new ideas within physics, related philosophical issues, or broader topics such as science and society

Briefs are characterized by fast, global electronic dissemination, straightforward publishing agreements, easy-to-use manuscript preparation and formatting guidelines, and expedited production schedules. We aim for publication 8–12 weeks after acceptance.

More information about this series at http://www.springer.com/series/8902

Peter Collas · David Klein

The Dirac Equation in Curved Spacetime

A Guide for Calculations

 Springer

Peter Collas
Department of Physics and Astronomy
California State University, Northridge
Northridge, CA, USA

David Klein
Department of Mathematics
and Interdisciplinary Research
Institute for the Sciences
California State University, Northridge
Northridge, CA, USA

ISSN 2191-5423　　　　　ISSN 2191-5431　(electronic)
SpringerBriefs in Physics
ISBN 978-3-030-14824-9　　　ISBN 978-3-030-14825-6　(eBook)
https://doi.org/10.1007/978-3-030-14825-6

Library of Congress Control Number: 2019933381

This Springer imprint is published by the registered company Springer Nature Switzerland AG
The registered company address is: Gewerbestrasse 11, 6330 Cham, Switzerland

Contents

Chapter 1
Introduction

The Dirac equation plays a fundamental role in relativistic quantum mechanics and in quantum field theory. It describes spin 1/2 particles, including electrons, neutrinos, muons, protons, neutrons, quarks, and their corresponding anti-particles. The Dirac equation has been extremely successful, even in its one-particle interpretation, in calculating the relativistic hydrogen atom spectrum, the g_s-factor of the electron's magnetic moment [1], and the spin-orbit coupling for the electron. It has also been used to calculate the Coulomb scattering amplitude [2] and even to obtain meaningful results in its ultra-relativistic limit, where the mass $m \rightarrow 0$ [3]. In fact a vast and rather recent research area has arisen, where the $(2 + 1)$ ultra-relativistic $(m = 0)$ Dirac equation is used to describe curved graphene and semi-metals [4].

Despite these successes, the Dirac equation's single-particle interpretation is not strictly possible since it describes both particles and antiparticles. Ultimately the only consistent approach is to regard the spinor ψ as a field. Additional problems arise when one is attempting to localize the particle to within a distance of the order of its Compton wavelength or when interacting with strong fields. Good reviews of these problems are given in [5–7]. Nevertheless when used with care the Dirac equation does give valuable insights and results.

In this book in the SpringerBriefs Series in Physics, we consider the Dirac equation in the context of general relativistic quantum mechanics. We assume as prerequisites familiarity with general relativity along with an exposure to the Dirac equation at the level of special relativistic quantum mechanics, although we also provide a review of this latter topic in our first section as a reference and framework for physical interpretations for the calculations and topics that follow.

Studies of the Dirac equation and its solutions, within the context of general relativity, go as far back as the 1940s [8]. Since then this area of research flourished as physicists and mathematicians found and explored exact and approximate solutions in several stationary [9–12], and cosmological backgrounds [13, 14], (*a more complete set of references can be found the relevant sections in the text below*). In Refs. [9, 10], Chandrasekhar separated and solved the Dirac equation in the Kerr spacetime

© The Author(s), under exclusive license to Springer Nature Switzerland AG 2019
P. Collas and D. Klein, *The Dirac Equation in Curved Spacetime*,
SpringerBriefs in Physics, https://doi.org/10.1007/978-3-030-14825-6_1

and showed the absence of superradiance for massive Dirac particles. In Ref. [11], Parker worked out in detail the curvature-induced energy level shifts in the relativistic hydrogen atom using perturbation theory and approximate Fermi coordinates. Parker and Pimentel [12], further extended and applied these results to hydrogen atoms in radial and circular orbits in the Schwarzschild spacetime. Audretsch and Schäffer in [13, 14], considered the hydrogen energy spectrum in expanding universes. In part because of the difficulties involved but also to gain insight on the effect of dimensionality on the solutions, researchers explored the Dirac equation in reduced, $(1 + 1)$, [15] and $(2 + 1)$, [16] spacetimes. In Ref. [15], Mann, Morsink, Sikkema and Steele, examined the solutions of both the Klein-Gordon and the Dirac equations in $(1 + 1)$ spacetimes containing a naked singularity, or a black hole. In both cases they found that the existence of a nontrivial curvature gave rise to a quantization condition for the energies of the solutions. More recently Sucu and Ünal [16] investigated aspects of the solutions of the Dirac equation in a $(2 + 1)$ curved cosmological spacetime involving a universe contracting from an infinite size to nonvanishing minimum radius and then expanding again to infinite size.

Within the scientific literature in all of the above areas, there is a potentially confusing array of conventions and methods. We outline the various approaches used in calculations involving the Dirac equation in curved spacetime, and clarify the subject by carefully pointing out the various conventions used, describing how they are related to each other, and by including examples from textbooks and articles, as well as examples of our own. In addition, some background material has been included in the appendices. It is our hope that graduate students and other researchers will find this book useful.

References

1. L.H. Ryder, *Quantum Field Theory*, 2nd edn. (Cambridge U. Press, New York, 1996), pp. 436–7
2. B. Thaller, V. Enss, Asymptotic observables and Coulomb scattering for the Dirac equation. Ann. de l' Institut Henri Poincaré A **45**, 147–171 (1986)
3. K. Konno, M. Kasai, General relativistic effects of gravity in quantum mechanics, a case of ultra-relativistic, spin 1/2 particles. Prog. Theor. Phys. **100**, 1145–1157 (1998)
4. A. Iorio, Curved spacetimes and curved graphene: a status report of the Weyl symmetry approach. Int. J. Mod. Phys. D **24**(1530013), 1–63 (2015)
5. C. Itzykson, J.-B. Zuber, *Quantum Field Theory* (McGraw-Hill Inc., New York, 1980)
6. B. Thaller, *The Dirac Equation* (Springer, Berlin, 1992)
7. W. Greiner, B. Müller, J. Rafelski, *Quantum Electrodynamics of Strong Fields* (Springer, New York, 1985)
8. E. Schrödinger, Maxwell's and Dirac's equations in the expanding universe. Proc. R. Irish Acad. A. **46**, 25–47 (1940)
9. S. Chandrasekhar, The solution of Dirac's equation in Kerr geometry. Proc. R. Soc. Lond. A. **349**, 571–575 (1976)
10. S. Chandrasekhar, *The Mathematical Theory of Black Holes* (Clarendon Press, Oxford, 1992)
11. L. Parker, One-electron atom as a probe of spacetime curvature. Phys. Rev. D **22**, 1922–1934 (1980)

12. L. Parker, L.O. Pimentel, Gravitational perturbation of the hydrogen spectrum. Phys. Rev. D **25**, 3180–3190 (1982)
13. J. Audretsch, G. Schäfer, Quantum mechanics of electromagnetically bounded spin-$\frac{1}{2}$ particles in an expanding universe: I. Influence of the expansion. Gen. Relativ. Gravit. **9**, 243–255 (1978)
14. J. Audretsch, G. Schäfer, Quantum mechanics of electromagnetically bounded spin-$\frac{1}{2}$ particles in expanding universes: II. Energy spectrum of the hydrogen atom. Gen. Relativ. Gravit. **9**, 489–500 (1978)
15. R.B. Mann, S.M. Morsink, A.E. Sikkema, T.G. Steele, Semiclassical gravity in $(1+1)$ dimensions. Phys. Rev. D **43**, 3948–3957 (1991)
16. Y. Sucu, N. Ünal, Exact solution of the Dirac equation in $2+1$ dimensional gravity. J. Math Phys. **48**, 052503 (2007)

Chapter 2
The Dirac Equation in Special Relativity

2.1 Review of the Dirac Equation

The purpose of this chapter is to review the Dirac equation in Minkowski spacetime and familiarize the reader with some of our notational conventions. This preliminary material may skipped by readers already familiar with its contents, and referred to as needed.

We begin with some notational conventions. We use the metric signature (-2), i.e., $(+, -, -, -)$, and adopt units so that $c = \hbar = 1$. The Minkowski metric is then expressed as

$$ds^2 = \eta_{\mu\nu}dx^\mu dx^\nu = dt^2 - dx^2 - dy^2 - dz^2 \,, \tag{2.1}$$

where μ and ν run over $(0, 1, 2, 3)$ or (t, x, y, z). We write $p = \left(p^0, \boldsymbol{p}\right)$, where,

$$p^0 = p_0 \,, \tag{2.2}$$

$$p^j = -p_j \,, \quad j = (1, 2, 3) = (x, y, z), \tag{2.3}$$

and

$$p^0 = p_0 = i\partial_t \,, \tag{2.4}$$

$$p^j = -p_j = -i\partial_j \,. \tag{2.5}$$

The *canonical momenta*, p_μ, are the 1-forms obtained by taking the derivatives of the Lagrangian, determined by the metric, with the respect to the generalized velocities, namely,

$$p_\mu = \frac{\partial L}{\partial \dot{x}^\mu} = \frac{1}{2} \frac{\partial}{\partial \dot{x}^\mu} \left(g_{\mu\nu}\dot{x}^\mu \dot{x}^\nu\right). \tag{2.6}$$

In Eq. (2.6) above we used $g_{\mu\nu}$ for the metric since this definition holds also in general relativity.

The Dirac equation for a free, spin $1/2$, particle of mass m in Minkowski spacetime is usually written as

$$i\gamma^\mu\partial_\mu\psi - m\psi = 0. \tag{2.7}$$

Introducing the Feynman "slash" notation,

$$\slashed{p} = \gamma^\mu p_\mu, \tag{2.8}$$

the Dirac Equation (2.7) can be expressed in a shorthand version as,

$$(\slashed{p} - mI)\psi = 0, \tag{2.9}$$

or more explicitly as,

$$i\gamma^0\partial_t\psi + i\gamma^1\partial_x\psi + i\gamma^2\partial_y\psi + i\gamma^3\partial_z\psi - m\psi = 0. \tag{2.10}$$

Here the solution ψ to the Dirac equation, is a 4-component spinor, so that,

$$\psi = \begin{pmatrix} \psi_1 \\ \psi_2 \\ \psi_3 \\ \psi_4 \end{pmatrix}. \tag{2.11}$$

The γ matrices are not uniquely specified and a variety of choices for them can be made, as we describe in Appendix B.

2.2   Plane Wave Solutions in the Standard Representation

We next consider plane wave solutions of the Dirac equation, first using the *standard representation* of the γ matrices (see Appendix B), then in Appendix D we show how this set of solutions can be transformed into the corresponding set in the *chiral* representation of the γ matrices. We note that the equations below regarding the energy projection operators and spin operators are representation independent. Among the most useful references for much of the material below are [1–5].

In general, finding solutions of the Dirac equation requires finding solutions to a set of coupled partial differential equations. For example, in the standard representation of the γ matrices, the Dirac equation (2.10), for the ψ_i of Eq. (2.11) becomes the following set of coupled partial differential equations:

$$i\partial_t\psi_1 + i\partial_z\psi_3 + i\partial_x\psi_4 + \partial_y\psi_4 - m\psi_1 = 0, \tag{2.12}$$

$$i\partial_t\psi_2 + i\partial_x\psi_3 - \partial_y\psi_3 - i\partial_z\psi_4 - m\psi_2 = 0, \tag{2.13}$$

$$-i\partial_z\psi_1 - i\partial_x\psi_2 - \partial_y\psi_2 - i\partial_t\psi_3 - m\psi_3 = 0, \tag{2.14}$$

$$-i\partial_x\psi_1 + \partial_y\psi_1 + i\partial_z\psi_2 - i\partial_t\psi_4 - m\psi_4 = 0. \tag{2.15}$$

However, as we show below it is sometimes possible find solutions to the Dirac equation by solving homogeneous partial differential equations, a substantial simplification.

We shall use the *Feynman-Stückelberg* interpretation for the particle-antiparticle states [6]. In this picture, negative energy particles are viewed as positive energy antiparticles moving backwards in time and with momenta in the opposite direction, see Eqs. (2.33), (2.34).

Remark 1 In general if the metric of the spacetime under consideration does not depend on one or more coordinates, say x^μ, then the corresponding canonical momenta p_μ are constants of the motion. This is true both in general relativity and quantum mechanics. In quantum mechanics the p_μ are the eigenvalues of the operators $i\partial_\mu$ and ψ is an eigenfunction of the canonical momenta p_μ. Thus, we can always take advantage of this fact by writing the dependence on these coordinates and the corresponding canonical momenta in an exponential overall factor.

In Minkowski spacetime the metric does not depend on any of the coordinates. Consequently, in accordance with Remark 1, our four-component spinors, ψ, will have the form of Eqs. (2.16) and (2.17) below.

$$\psi^{(+)} = u(p)\, e^{-ip_\mu x^\mu}, \tag{2.16}$$

$$\psi^{(-)} = v(p)\, e^{ip_\mu x^\mu}, \tag{2.17}$$

where $p_t = p^t \geq 0$ and $\psi^{(+)}$ will denote the positive energy solutions and $\psi^{(-)}$ the negative energy solutions.

Substituting Eqs. (2.16) and (2.17) in Eq. (2.7), we find the set of algebraic equations below,

$$(\not{p} - mI)u(p) = 0, \tag{2.18}$$

$$(\not{p} + mI)v(p) = 0. \tag{2.19}$$

Observe that Eqs. (2.18) and (2.19) are systems of homogeneous equations for the components of $u(p)$ and $v(p)$. These systems will have a non-trivial solutions only if

$$\det(\not{p} \pm mI) = 0. \tag{2.20}$$

Equation (2.20) gives us the (representation independent) condition

$$\left(p^2 - m^2\right)^2 = 0, \tag{2.21}$$

where $p^2 = \left(p^t\right)^2 - \boldsymbol{p}^2 \equiv E^2 - \boldsymbol{p}^2$, and therefore Eq. (2.21) may be rewritten as

$$\left(p^2 - m^2\right) = \left[\left(E - \sqrt{\boldsymbol{p}^2 + m^2}\right)\left(E + \sqrt{\boldsymbol{p}^2 + m^2}\right)\right] = 0. \tag{2.22}$$

We see that condition (2.20) is satisfied for both $E = \pm\sqrt{\boldsymbol{p}^2 + m^2}$. Our set of solutions consists of four linearly independent 4-component spinors. In this section we will use ψ for the spinors in the standard representation and ϕ for the spinors in the chiral representation (Appendix B). We also emphasize that

$$p_t = +\sqrt{(p_x)^2 + (p_y)^2 + (p_z)^2 + m^2}. \tag{2.23}$$

It is easy to verify that the two positive energy spinors $u^{(1)}(p)$ and $u^{(2)}(p)$ below, satisfy Eq. (2.18). The superscripts (1), (2) refer to the spin orientations in the positive and negative z-directions, $S_z = +1/2$ and $S_z = -1/2$, respectively, discussed below. The factor, $N = N(p)$, is an arbitrary normalization that will be dealt with in Appendix D.3.

$$u^{(1)}(p) = N \begin{pmatrix} 1 \\ 0 \\ -\dfrac{p_z}{p_t + m} \\ -\dfrac{p_x + ip_y}{p_t + m} \end{pmatrix}, \quad u^{(2)}(p) = N \begin{pmatrix} 0 \\ 1 \\ \dfrac{-p_x + ip_y}{p_t + m} \\ \dfrac{p_z}{p_t + m} \end{pmatrix}. \tag{2.24}$$

Using Eq. (2.19), we obtain the two negative energy spinors $v^{(1)}(p)$ and $v^{(2)}(p)$ below,

$$v^{(1)}(p) = N \begin{pmatrix} -\dfrac{p_z}{p_t + m} \\ -\dfrac{p_x + ip_y}{p_t + m} \\ 1 \\ 0 \end{pmatrix}, \quad v^{(2)}(p) = N \begin{pmatrix} \dfrac{-p_x + ip_y}{p_t + m} \\ \dfrac{p_z}{p_t + m} \\ 0 \\ 1 \end{pmatrix}. \tag{2.25}$$

The four solutions above are, are of course, linearly independent. We summarize here some notation and results. To avoid notational ambiguities we will write, for example,

$$\psi^{(+)(\alpha)}(x) = \begin{pmatrix} u_1^{(\alpha)}(p) \\ u_2^{(\alpha)}(p) \\ u_3^{(\alpha)}(p) \\ u_4^{(\alpha)}(p) \end{pmatrix} e^{-ip_\mu x^\mu}, \tag{2.26}$$

and so on, thus,

$$\psi^{(+)(\alpha)}(x) = u^{(\alpha)}(p)e^{-ip_\mu x^\mu}, \tag{2.27}$$

$$\psi^{(-)(\alpha)}(x) = v^{(\alpha)}(p)e^{ip_\mu x^\mu}, \quad \alpha = (1, 2). \tag{2.28}$$

Each member of the set ψ is an eigenstate of the energy, and momentum. We have adopted the notation below so that Eqs. (2.16) and (2.17) are:

$$\left. \begin{array}{l} \psi^{(+)(1)}, \quad S_z = +\frac{1}{2} \\ \psi^{(+)(2)}, \quad S_z = -\frac{1}{2} \end{array} \right\} \quad \text{positive energy} \tag{2.29}$$

$$\left. \begin{array}{l} \psi^{(-)(1)}, \quad S_z = +\frac{1}{2} \\ \psi^{(-)(2)}, \quad S_z = -\frac{1}{2} \end{array} \right\} \quad \text{negative energy} \tag{2.30}$$

where as noted S_z is the spin the $z-$direction as we explain in the next section.

2.3 The Hamiltonian, Energy and Spin Operators

A short calculation using Eqs. (2.27) and (2.28) shows that,

$$i\partial_t \psi^{(+)(\alpha)}(x) = p_t \, \psi^{(+)(\alpha)}(x) = p^t \, \psi^{(+)(\alpha)}(x), \tag{2.31}$$

$$i\partial_t \psi^{(-)(\alpha)}(x) = -p_t \, \psi^{(-)(\alpha)}(x) = -p^t \, \psi^{(-)(\alpha)}(x), \tag{2.32}$$

where $p_t = p^t = E > 0$, and similarly for the momenta we have, for example,

$$-i\partial_z \psi^{(+)(\alpha)}(x) = -p_z \, \psi^{(+)(\alpha)}(x) = p^z \, \psi^{(+)(\alpha)}(x), \tag{2.33}$$

$$-i\partial_z \psi^{(-)(\alpha)}(x) = +p_z \, \psi^{(-)(\alpha)}(x) = -p^z \, \psi^{(-)(\alpha)}(x). \tag{2.34}$$

Remark 2 It is important to note that the (eigenvalues) of the components of the 3-momenta in Eq. (2.34), have opposite signs from the ones in Eq. (2.33). We make use of this fact below in connection with the Hamiltonian.

Eqs (2.31) and (2.32) suggest that the $\psi^{(\pm)(\alpha)}(x)$ are energy eigenstates for an appropriate Dirac Hamiltonian H satisfying the eigenvalue equation,

$$i\partial_t \psi = H\psi = E\psi. \tag{2.35}$$

The Hamiltonian H (for Minkowski spacetime) is constructed by multiplying Eq. (2.7) or (2.10) on the left by γ^0 and finding $i\partial_t\psi$, thus,

$$H = \gamma^0 \left(-\gamma^k p_k + mI\right), \quad k = (1, 2, 3) = (x, y, z). \tag{2.36}$$

The expression above for H is representation independent. Some authors write the Hamiltonian in terms of the Dirac matrices, α and β, where $\gamma^k = \gamma^0\alpha^k$ and $\gamma^0 = \beta$. Using the positive energy spinors, $u^{(+)(\alpha)}$, Eq. (2.27), (which are in the Dirac representation), we verify that

$$H\psi^{(+)(\alpha)} = p_t\psi^{(+)(\alpha)}. \tag{2.37}$$

In light of Remark 2 above, and in order to get the correct result for the negative energy solutions, we have to apply the Hamiltonian H on the spinors, Eq. (2.28), *after we change the signs of all the space components, $p_k \to -p_k$,* (e.g., see Peskin and Schroeder [2], p. 53 or Itzykson and Zuber [1], p. 58). Then we obtain

$$H\psi^{(-)(\alpha)} = -p_t\psi^{(-)(\alpha)}, \tag{2.38}$$

in agreement with Eq. (2.32), where now

$$\psi^{(-)(\alpha)} = v^{(\alpha)}(p_t, -p_k)e^{i\left(p_t t - p_k x^k\right)}. \tag{2.39}$$

Alternatively one can introduce two Hamiltonians, $H_{(\pm)}$, where H_+, is identical to the Hamiltonian of Eq. (2.36) above, while in $H_{(-)}$, for the negative energy solutions, we reverse the signs of the p_k, in Eq. (2.36), so that now

$$H_{(-)} = \gamma^0 \left(\gamma^k p_k + mI\right), \quad k = (1, 2, 3) = (x, y, z), \tag{2.40}$$

and

$$H_{(-)}\psi^{(-)(\alpha)} = -p_t\psi^{(+)(\alpha)}, \tag{2.41}$$

where now $\psi^{(-)(\alpha)}$ is the one given by Eq. (2.28).

Finally, we introduce the normalized energy projection operators,

$$\Lambda_{\pm} = \frac{\pm \not{p} + mI}{2m}. \tag{2.42}$$

One can easily deduce, using the solutions in Eqs. (2.24) and (2.25), that

$$\Lambda_{+} u^{(1)}(p) = u^{(1)}(p), \tag{2.43}$$

$$\Lambda_{+} u^{(2)}(p) = u^{(2)}(p), \tag{2.44}$$

$$\Lambda_{+} v^{(1)}(p) = \Lambda_{+} v^{(2)}(p) = 0, \tag{2.45}$$

$$\Lambda_{-} v^{(1)}(p) = v^{(1)}(p), \tag{2.46}$$

$$\Lambda_{-} v^{(2)}(p) = v^{(2)}(p), \tag{2.47}$$

$$\Lambda_{-} u^{(1)}(p) = \Lambda_{-} u^{(2)}(p) = 0. \tag{2.48}$$

For a rigorous mathematical discussion of the Dirac Hamiltonian and related matters one may consult Thaller [7], in particular Sect. 1.4.1–1.4.6.

We turn now to a discussion of the spin operators *in the particle rest frame* along the three axes directions (see Ohlsson [3] for more details). Although one can choose the spin orientation axis in an arbitrary direction, it is far simpler and more useful to choose it along the particle's 3-momentum $\boldsymbol{p}$. This is sometimes referred to as a *helicity* basis. Define,

$$\Sigma^{k} = \gamma^{5}\gamma^{0}\gamma^{k}, \quad k = (1, 2, 3) = (x, y, z). \tag{2.49}$$

Then, for example, if ψ is an eigenstate of spin in the z-direction, as indicated for the spinors in Eqs. (2.29) and (2.30) (see Eqs. (2.24) and (2.25)) we should find that

$$\Sigma^{3}\psi = \psi, \tag{2.50}$$

and a short calculation yields the above result. Strictly speaking one should define the operators,

$$S^{k} = \tfrac{1}{2} \Sigma^{k} \tag{2.51}$$

in order to obtain the correct eigenvalue of $\frac{1}{2}$, but we omit this factor for convenience.

We consider the related *spin projection operators* again restricting ourselves to the particle rest frame but allowing for arbitrary polarization directions. The covariant spin operator is given by,

$$\Sigma = \left(0, \Sigma^{1}, \Sigma^{2}, \Sigma^{3}\right), \tag{2.52}$$

and so we may introduce the normalized polarization vector

$$s = \left(0, s^{x}, s^{y}, s^{z}\right), \quad s^{2} = \eta_{jk}s^{j}s^{k} = -1, \tag{2.53}$$

along with the corresponding 1-form

$$\tilde{s} = \left(0, s_x, s_y, s_z\right) = \left(0, -s^x, -s^y, -s^z\right). \tag{2.54}$$

For example for polarization in the positive z-direction, we let

$$s = (0, 0, 0, 1), \quad \tilde{s} = (0, 0, 0, -1). \tag{2.55}$$

We can now write the *spin projection operators*

$$S_\pm = \frac{1}{2}\left[I \pm \left(-s_\mu \Sigma^\mu\right)\right]. \tag{2.56}$$

One can easily verify that

$$S_\pm u^{(i)} = u^{(i)}, \quad S_\pm v^{(i)} = v^{(i)}, \tag{2.57}$$

where in the superscripts of the u and v the $(i) = (1, 2)$ for polarization in the positive and negative directions respectively.

Finally we display the Foldy-Wouthuysen spin operator [8] which can be used if the particle is not necessarily in the rest frame:

$$\Sigma^k = \frac{\gamma^5 \gamma^0}{p_t}\left[m\gamma^k + \left(p_j \gamma^j\right)\gamma^k + \frac{p_k}{p_t + m}\left(p_j \gamma^j\right) + p_k I\right], \tag{2.58}$$

where,

$$p_j \gamma^j = p_x \gamma^1 + p_y \gamma^2 + p_z \gamma^3. \tag{2.59}$$

Equation (2.58) does, of course, simplify to Eq. (2.49) when $p_x = p_y = p_z = 0$, $p_t = m$.

Appendix D provides some additional topics of importance for the Dirac equation in special relativity, including a statements of Pauli's fundamental theorem relating sets of γ matrices in different representations, plane wave solutions in the chiral representation, and a discussion of the normalization of the plane wave solutions $\psi^{(\pm)(\alpha)}$. In particular in Sect. D.2, we work out an example of the details of how the spin polarization is calculated for $u(p)$ in the chiral representation.

We conclude this chapter with the warning that although suitable Dirac Hamiltonians can be written in curved spacetimes [9, 10], the same cannot be said for the spin operators [11].

References

1. C. Itzykson, J.-B. Zuber, *Quantum Field Theory* (McGraw-Hill Inc., New York, 1980)
2. M.E. Peskin, D.V. Schroeder, *An Introduction to Quantum Field Theory*. (Perseus Books, Reading, 1995)
3. T. Ohlsson, *Relativistic Quantum Mechanics* (Cambridge U. Press, Cambridge, 2011)
4. M.A. Thomson, Particle Physics. Handout 2: The Dirac Equation (2011), http://www.hep.phy.cam.ac.uk/~thomson/lectures/lectures.html
5. W. Greiner, *Relativistic Quantum Mechanics, Wave Equations*, 3rd edn. (Springer, New York, 2000)
6. A. Wachter, *Relativistic Quantum Mechanics* (Springer, Berlin, 2011)
7. B. Thaller, *The Dirac Equation* (Springer-Verlag, Berlin, 1992)
8. P. Caban, J. Rembieliński, M. Włodarczyk, Spin operator in the Dirac theory. Phys. Rev. A **88**, 022119 (2013)
9. L. Parker, One-electron atom as a probe of spacetime curvature. Phys. Rev. D **22**, 1922–1934 (1980)
10. X. Huang, L. Parker, Hermiticity of the Dirac Hamiltonian in curved spacetime. Phys. Rev. D **79**, 024020 (2009)
11. L.H. Ryder, Spin in special and general relativity, in *Gyros, Clocks, Interferometers ...* ed by C. Lämmerzahl, C.W.F. Everitt, F.W. Hehl. Testing Relativistic Gravity in Space, (Springer, Berlin, 2001)

Chapter 3
The Spinorial Covariant Derivative

3.1 The Fock-Ivanenko Coefficients

In the calculations that follow we will specify our metric signature as needed. We adopt the convention that upper case latin indices run over $(0, 1, 2, 3)$ while greek indices are coordinate indices, e.g., (t, r, θ, ϕ). Also we shall adopt "Planck units" so that, $c = \hbar = G = 1$, [1]. In such units all physical quantities and variables are dimensionless.

In order to accommodate spinors in general relativity we need the tetrad formalism. The tetrad formalism is reviewed in Appendix A, where apart from the standard material we have included some additional material relevant to spinors [2].

From the basic tetrad expression, Eq. (A.4) in Appendix A, namely,

$$\eta_{AB} = e_A{}^\alpha e_B{}^\beta g_{\alpha\beta}, \tag{3.1}$$

we see that $\eta_{00} = e_0{}^\alpha e_0{}^\beta g_{\alpha\beta}$, and thus by definition the tetrad vector e_0 is a velocity field at least momentarily tangent to a timelike path. This is what Schutz [3] refers to as the "momentarily comoving reference frame" (MCRF) and it is in this sense that our choice of a tetrad vector set, $e_A{}^\alpha$, determines the frame we shall refer to as the *the reference frame for the Dirac particle*, or *the particle frame* for short.

For example in reference [4], Sect. VI, Parker considers a freely falling hydrogenic atom, with its nucleus on the geodesic (an approximation), and constructs approximate Fermi coordinates along the chosen geodesic. The corresponding tetrad is referred to as *the proper frame*. In general the choice of a tetrad set may be dictated for reasons of convenience and one should read the comments in Remark 11 and keep in mind the fact that from a given tetrad one can obtain an infinity of tetrads related to each other by local Lorentz transformations (see Sect. 5.2).

In general relativity the spinors, $\psi(x)$, are sections of the spinor bundle. We limit ourselves to presenting the bare essentials required for calculations, and on clarifying the different sign conventions related to the definition of the spinorial covariant derivative, the spinor affine connection, Γ_μ, and Fock-Ivanenko coefficients Γ_C.

Each component of a spinor transforms as a scalar function under general coordinate transformations, so this kind of transformation is straightforward. However, the transformation of spinors under tetrad rotations requires additional formalism.

If we change from an initial set of tetrad vector fields, h_A, to another set, e_A, then the new tetrad vectors can be expressed as linear combinations of the old as shown below

$$e_A{}^\mu = \Lambda_A{}^B h_B{}^\mu \, , \tag{3.2}$$

We show in Appendix A.4 that Λ is a Lorentz matrix. So in the context of general relativity *the Lorentz group is the group of tetrad rotations* [2]. We also remark that the Λ matrices are in general spacetime-dependent and we refer to them as local Lorentz transformations.

In order to write the Dirac equation in general relativity, we also need to introduce the spacetime dependent matrices $\bar{\gamma}^\alpha(x)$. The $\bar{\gamma}^\alpha$ matrices are related to the constant special relativity gamma matrices, γ^A, by the relation

$$\bar{\gamma}^\alpha(x) := e_A{}^\alpha(x)\gamma^A \, , \tag{3.3}$$

Using Eq. A.4 we can now relate the anti-commutators below,

$$\{\gamma^A, \gamma^B\} = \varepsilon\, 2\eta^{AB} I \, , \tag{3.4}$$

$$\{\bar{\gamma}^\alpha(x), \bar{\gamma}^\beta(x)\} = \varepsilon\, 2g^{\alpha\beta} I \, , \tag{3.5}$$

where $\varepsilon = \pm 1$. We note that the matrices in Eq. (3.23) anticommute for $\alpha \neq \beta$, only if the metric is diagonal.

A spinor, ψ, may be defined as a quantity that transforms as

$$\tilde{\psi}_e = L\, \psi_h \, , \tag{3.6}$$

where $L = L(x)$ is the spacetime-dependent spinor representative of a tetrad rotation $\Lambda = \Lambda(x)$ [2]. We initially follow the sign conventions of references [5–7], and although all these references use the metric signature $(+, -, -, -)$, we shall maintain, wherever possible, greater generality.

The derivative of a spinor does not transform like a spinor since

$$\tilde{\psi}_{,\mu} = L\, \psi_{,\mu} + L_{,\mu}\, \psi \, . \tag{3.7}$$

Therefore we define the covariant derivative of a spinor by the expression,

$$D_\mu \psi = I \psi_{,\mu} + \Gamma_\mu \psi \, , \tag{3.8}$$

with the spinor affine connection, Γ_μ to be determined. The connection Γ_μ is a matrix, actually four matrices, that is, $(\Gamma_\mu)_a{}^b$. We require that,

$$\tilde{D}_\mu \tilde{\psi} = L D_\mu \psi \,, \tag{3.9}$$

where

$$\tilde{D}_\mu \tilde{\psi} = I \tilde{\psi},_\mu + \tilde{\Gamma}_\mu \tilde{\psi} \,. \tag{3.10}$$

Equation (3.9) is satisfied if we let

$$\tilde{\Gamma}_\mu = L \, \Gamma_\mu L^{-1} - L,_\mu L^{-1} \,, \tag{3.11}$$

since then

$$\tilde{D}_\mu \tilde{\psi} = \partial_\mu (L\psi) + \tilde{\Gamma}_\mu \tilde{\psi} \,, \tag{3.12}$$

$$= (L,_\mu)\psi + L\psi,_\mu + L\Gamma_\mu \psi - (L,_\mu)\psi \,. \tag{3.13}$$

$$= L(I\psi,_\mu + \Gamma_\mu \psi) \,. \tag{3.14}$$

With a slight abuse of notation we write the spinor covariant derivative acting on a spinor $\psi(x)$ as

$$D_\mu \psi = \left(I\partial_\mu + \Gamma_\mu \right) \psi := \left(\partial_\mu + \Gamma_\mu \right) \psi \,, \tag{3.15}$$

where we may omit the identity matrix factor I in the second part of Eq. (3.15). We now proceed to deduce the expression for Γ_μ.

Under the assumption that the operator D_μ is a connection and therefore a derivation (i.e., satisfies the product rule for tensor products), it may be extended as an operator on a matrix-valued field M, [7, 8]. By writing M as a linear combination of tensor products of vectors with co-vectors, a calculation shows that

$$D_\mu M = \nabla_\mu M + \left[\Gamma_\mu, M \right] \,. \tag{3.16}$$

In particular, if $M = I$, then $D_\mu M = 0$. We now impose the additional requirement that the derivative, D_μ, is *metric compatible*, i.e.,

$$D_\mu g^{\alpha\beta} I = 0 \,, \tag{3.17}$$

where $g^{\alpha\beta}$ in this expression is understood to be a scalar (the element of a matrix) rather than a matrix. Recalling Eq. (3.23),

$$\varepsilon \, 2g^{\alpha\beta} I = \{\bar{\gamma}^\alpha(x), \bar{\gamma}^\beta(x)\} \,, \tag{3.18}$$

we see that Eq. (3.17) is equivalent to

$$D_\mu \left(\{\bar{\gamma}^\alpha(x), \bar{\gamma}^\beta(x)\} \right) = 0 \,, \tag{3.19}$$

and a *sufficient* condition for the above equation is

$$D_\mu \bar{\gamma}^\nu(x) = 0 . \tag{3.20}$$

The operator D_μ of Eq. (3.16) acting on $\bar{\gamma}^\nu$ is,

$$D_\mu \bar{\gamma}^\nu = \nabla_\mu \bar{\gamma}^\nu + \left[\Gamma_\mu, \bar{\gamma}^\nu \right] . \tag{3.21}$$

Thus Eq. (3.20) is

$$D_\mu \bar{\gamma}^\nu = \bar{\gamma}^\nu{}_{,\mu} + \Gamma^\nu{}_{\lambda\mu} \bar{\gamma}^\lambda + \Gamma_\mu \bar{\gamma}^\nu - \bar{\gamma}^\nu \Gamma_\mu = 0 . \tag{3.22}$$

We refer the reader to Appendix C, for further details of the effect of signature choices on the Γ_μ and the Dirac equation.

Remark 3 Reference [2], Eq. (13.27), for example, has a $(-)$ sign in front of the commutator in Eq. (3.21) thus, effectively, changing the sign of Γ_μ. This is compensated for by writing $\left(\partial_\mu - \Gamma_\mu \right)$ in Eq. (3.15).

We now introduce the *spin* connection (coefficients) $\omega^A{}_{B\mu}$ by the relation below (e.g., see [9], pp. 222–224, and [10], p. 487),

$$\omega^A{}_{B\mu} := -e_B{}^\nu \left(\partial_\mu e^A{}_\nu - \Gamma^\lambda{}_{\mu\nu} e^A{}_\lambda \right) . \tag{3.23}$$

$$\omega_{AB\mu} = e_{A\beta} \nabla_\mu e_B{}^\beta = g_{\beta\alpha} e_A{}^\alpha \nabla_\mu e_B{}^\beta = \eta_{AC} e^C{}_\beta \nabla_\mu e_B{}^\beta . \tag{3.24}$$

We see from the second (or third) equality in Eq. (3.24) that the metric signature will affect the signs of the $\omega_{AB\mu}$.

Exercise 1 *Obtain Eq. (3.23) using Eq. (3.24) and the result of Proposition 1 below. (Hint: Pay attention to the ordering of the indices A and B in Eqs. (3.23) and (3.24).)*

Proposition 1 *The $\omega_{AB\mu}$ are antisymmetric in A and B.*

Proof Recall Eq. (A.4), then we have that (cf. [10], p. 489),

$$\nabla_\mu \eta_{AB} = e_B{}^\beta \left(\nabla_\mu e_A{}^\alpha \right) g_{\alpha\beta} + e_A{}^\alpha \left(\nabla_\mu e_B{}^\beta \right) g_{\alpha\beta} = 0 ,$$
$$e_{B\alpha} \left(\nabla_\mu e_A{}^\alpha \right) + e_{A\beta} \left(\nabla_\mu e_B{}^\beta \right) = 0 ,$$
$$\omega_{BA\mu} + \omega_{AB\mu} = 0 . \tag{3.25}$$

Remark 4 The $\omega_{AB\mu}$ are also written as $\omega_{\mu AB}$, [9]. Lord refers to $\nabla_\mu e_B{}^\beta$ as the Ricci rotation coefficients while in our nomenclature the Ricci rotation coefficients are given by Eq. (3.68), namely their components along the tetrad field $e_C{}^\mu$. Our point here is that the terminology varies a little in the literature and care is required.

Using Eqs. (3.3) and (3.4) in Eq. (3.21), one can show that the Γ_μ below satisfies Eq. (3.22) and hence (3.17),

$$\Gamma_\mu = \frac{\varepsilon}{4}\,\omega_{AB\mu}\,\gamma^A\gamma^B = \frac{\varepsilon}{2}\,\omega_{AB\mu}\,\Sigma^{AB}\,, \tag{3.26}$$

where

$$\Sigma^{AB} = \frac{1}{4}\left[\gamma^A, \gamma^B\right]. \tag{3.27}$$

We have included an ε factor in the expression for Γ_μ in order take into account the choice made in Eq. (3.4). The factor of $1/4$ in Γ_μ, Eq. (3.26), compensates for the factor of 2 in Eq. (3.4) and thus is dimension independent. The reader may consult Fursaev and Vassilevich [11], p. 16, and Birrell and Davies [6] for different lines of reasoning.

Remark 5 We caution the reader that Parker and Toms [9] define a Γ_μ on p. 145 which is our Eq. (3.26) with $\varepsilon = -1$ and a B_μ on p. 228, which is our Eq. (3.26) with $\varepsilon = +1$. No problem arises since the corresponding Dirac equations also differ in the appropriate sign.

The *Fock-Ivanenko* coefficients, Γ_C, are given by

$$\Gamma_C = e_C{}^\mu\,\Gamma_\mu\,, \tag{3.28}$$

thus we may write

$$D_C\,\psi = (I e_C + \Gamma_C)\,\psi\,. \tag{3.29}$$

Finally, for a free spin $1/2$ particle of mass m we have the Dirac equation in curved spacetime,

$$i\gamma^C D_C\,\psi - m\psi = 0\,. \tag{3.30}$$

Using Eqs. (3.3), (3.15), and (3.28), we may also write Eq. (3.30) in the form

$$i\bar{\gamma}^\mu(\partial_\mu + \Gamma_\mu)\,\psi - m\psi = 0\,, \tag{3.31}$$

$$i\bar{\gamma}^\mu D_\mu\psi - m\psi = 0\,. \tag{3.32}$$

Remark 6 We note that the $e_C = e_C{}^\gamma\,\partial_\gamma$, in Eqs. (3.29), (3.30), is regarded as a differential operator, thus for our four component spinor ψ, we have, after re-inserting the identity matrix I,

$$e_C I\,\psi = \begin{pmatrix} e_C{}^\gamma\,\partial_\gamma\,\psi_1 \\ e_C{}^\gamma\,\partial_\gamma\,\psi_2 \\ e_C{}^\gamma\,\partial_\gamma\,\psi_3 \\ e_C{}^\gamma\,\partial_\gamma\,\psi_4 \end{pmatrix}. \tag{3.33}$$

Proposition 2 *If the tetrad $e_B{}^\beta$ in Eq. (3.24) is parallel along a timelike geodesic with tangent $e_0{}^\mu$, then it is a Fermi tetrad, (see Eq. (A.15)) and it turns out that the Fock-Ivanenko coefficient $\Gamma_0 = 0$.*

Proof Using Eqs. (3.24), (3.26), we obtain the expression below for Eq. (3.28),

$$\Gamma_C = \frac{\varepsilon}{4} e_{A\beta} \left(e_C{}^\mu \nabla_\mu e_B{}^\beta \right) \gamma^A \gamma^B , \qquad (3.34)$$

thus if $e_B{}^\beta$ is parallel, the terms in parentheses vanish for $C = 0$.

Remark 7 Conversely, if $\Gamma_0 \neq 0$, then $e_B{}^\beta$ is not parallel. However, it may happen that $\Gamma_0 = 0$, while $e_B{}^\beta$ is not parallel.

3.2   The Dirac Hamiltonian in Curved Spacetime

In order to find the Dirac Hamiltonian in curved spacetime, we follow similar steps to the ones we used in finding the Dirac Hamiltonian Eq. (2.36) in Sect. 2.3. In order to define the Dirac Hamiltonian it is necessary that the metric depends explicitly on a timelike 1-form dt. For that purpose we write our metric in the $(3 + 1)$ decomposition, where N and N^i are the *lapse and shift*, respectively, [12, 13]. Since here we are using signature (-2), the 3-metric $h_{ij} < 0$. Also N and N^i satisfy the inequalities: $N > 0$ and $N^2 + h_{ij} N^i N^j > 0$. Thus we write

$$ds^2 = N^2 dt^2 + h_{ij} \left(dx^i + N^i dt \right) \left(dx^j + N^j dt \right) = g_{\alpha\beta} dx^\alpha dx^\beta , \qquad (3.35)$$

In Eq. (3.35), $-h_{ij}$ is the induced Riemannian metric on the spacelike Cauchy hypersurfaces which are orthogonal to the vector ∂_t. Now we use Eq. (3.32) and multiply on the left by $\bar\gamma^t$. We choose $\varepsilon = +1$ in Eq. (3.5) and recall that

$$\left(\bar\gamma^t \right)^2 = g^{tt} I. \qquad (3.36)$$

We find

$$H = -\frac{i}{g^{tt}} \bar\gamma^t \bar\gamma^k \left(I \partial_k + \Gamma_k \right) - i \Gamma_t + \frac{m}{g^{tt}} \bar\gamma^t , \qquad (3.37)$$

where $k = (x^1, x^2, x^2)$.

Exercise 2 *Using Eq. (3.31), obtain Eq. (3.37).*

As Parker [4] pointed out, the above Hamiltonian is Hermitian for stationary metrics but is not Hermitian if the metric depends explicitly on time. It is true, of course, that the canonical momentum, p_t, is not conserved in this case. Nevertheless is is important, if possible, to define a Hermitian Hamiltonian valid, for example, for cosmological metrics since it would enable one to explore by means of perturbation theory effects on bound systems, such as molecules and atoms, that may result from the time dependence of the Riemann tensor. Also one would be able to calculate, using perturbation theory, shifts in energy levels and transition rates [14]. Huang and Parker [14] did find a correction term needed for the Dirac Hamiltonian to be Hermitian in all metrics.

3.3 Transformation Properties of the Spin Operator

In this subsection we return briefly to Minkowski spacetime (or an open submanifold of Minkowski spacetime) in order to discuss how the spin operators transform under tetrad rotations. We recall the definition of the spin operators, Eq. (2.49),

$$\Sigma^A = \gamma^5 \gamma^0 \gamma^A, \quad A = (1, 2, 3) = (x, y, z), \tag{3.38}$$

corresponding to polarization in the positive x, y, z, directions, in the particle's rest frame. We show below how the operator Σ^A transforms under a (local) field of Lorentz transformations Λ acting on the Minkowski tetrad field associated with the Minkowski coordinates (t, x, y, z), and we find two equivalent expressions for the corresponding transformation of the operator Σ^A.

The Σ^A of Eq. (3.38) are spin operators in the particle's rest frame provided that our tetrad vectors are canonical for Minkowski spacetime. We express the canonical tetrad field as:

$$e_0 = (1, 0, 0, 0) = \partial_t, \tag{3.39}$$
$$e_1 = (0, 1, 0, 0) = \partial_x, \tag{3.40}$$
$$e_2 = (0, 0, 1, 0) = \partial_y, \tag{3.41}$$
$$e_3 = (0, 0, 0, 1) = \partial_z. \tag{3.42}$$

The above tetrads satisfy the orthonormality relations,

$$e_A{}^\alpha e_B{}^\beta \eta_{\alpha\beta} = \eta_{AB}, \tag{3.43}$$

where η_{AB} is the Minkowski metric. If we now perform a local Lorentz transformation Λ on the tetrad vectors (3.39)–(3.42), we obtain a new set of tetrad vectors, $h_A{}^\alpha$, which are related to the ones above by the equation (written below in terms of the corresponding 1-forms)

$$h^A{}_\alpha = \Lambda^A{}_B \, e^B{}_\alpha, \tag{3.44}$$

where

$$\Lambda^A{}_C \Lambda^B{}_D \, \eta_{AB} = \eta_{CD}, \tag{3.45}$$

and, of course,

$$h^A{}_\alpha h^B{}_\beta \eta_{AB} = \eta_{\alpha\beta} . \tag{3.46}$$

The spacetime dependent γ matrices Eq. (3.3), are then given by,

$$\bar\gamma^\alpha(x) := h_A{}^\alpha(x)\gamma^A . \tag{3.47}$$

Using Eq. (3.47) and $h^B{}_\alpha h_A{}^\alpha = \delta^B{}_A$, we obtain

$$\gamma^B = h^B{}_\alpha \bar\gamma^\alpha . \tag{3.48}$$

It is easy to check that L and Λ satisfy the consistency relations [15]

$$L^{-1}\gamma^A L = \Lambda^A{}_B \gamma^B , \tag{3.49}$$

and

$$\gamma^0 L^\dagger \gamma^0 = L^{-1} , \tag{3.50}$$

Then with the matrix L related to Λ by Eq. (3.49), we get, by rearranging terms,

$$L \gamma^C L^{-1} = \gamma^B \Lambda_B{}^C , \tag{3.51}$$

$$= h^B{}_\alpha \Lambda_B{}^C \bar\gamma^\alpha. \tag{3.52}$$

But from Eq. (3.44), $h^B{}_\alpha = \Lambda^B{}_D e^D{}_\alpha$, so that,

$$L \gamma^C L^{-1} = \Lambda^B{}_D e^D{}_\alpha \Lambda_B{}^C \bar\gamma^\alpha, \tag{3.53}$$

$$= \delta_D{}^C e^D{}_\alpha \bar\gamma^\alpha. \tag{3.54}$$

$$= e^C{}_\alpha \bar\gamma^\alpha := \bar\gamma^C. \tag{3.55}$$

It now follows immediately that for each $A = x, y, z$,

$$\overline\Sigma^A = \bar\gamma^5 \bar\gamma^t \bar\gamma^A = (L\gamma^5 L^{-1})(L\gamma^t L^{-1})(L\gamma^A L^{-1}) = L\Sigma^A L^{-1}. \tag{3.56}$$

We summarize this results of this subsection in the form of the following theorem.

Theorem 1 *Let Ψ be a spinor defined in a region of Minkowski space with coordinates (t, x, y, z) with the canonical tetrad field given by Eqs. (3.39)–(3.42). Let Λ be a local Lorentz transformation and let $\bar\gamma^\alpha(x)$ be defined by Eqs. (3.47) and (3.44), and let $\overline\Sigma^A := \bar\gamma^5 \bar\gamma^t \bar\gamma^A$. Suppose that L and Λ satisfy (3.49) and (3.50). Then for $A = x, y$ or z,*

$$\Sigma^A \Psi = \Psi \quad \text{if and only if} \quad \overline\Sigma^A L\Psi = L\Psi. \tag{3.57}$$

3.4 The Ricci Rotation Coefficient Approach

We now give an alternative, and possibly more efficient, way of calculating the Fock-Ivanenko coefficients. We remark again that the terminology and definitions of some the quantities below vary in the literature and one has to be very careful. We work in the tetrad frame and define the *structure coefficients* (or *structure constants*), $C^D{}_{AB}$, by the relations below, [16–18].

$$de^D = -\frac{1}{2} C^D{}_{AB} \, e^A \wedge e^B . \tag{3.58}$$

An equivalent expression is

$$[e_A, e_B]^\gamma = e_A{}^\alpha \, \partial_\alpha \, e_B{}^\gamma - e_B{}^\beta \, \partial_\beta \, e_A{}^\gamma = C^D{}_{AB} \, e_D{}^\gamma , \tag{3.59}$$

while a most convenient expression is

$$C^D{}_{AB} = \left(e^D{}_{\alpha, \beta} - e^D{}_{\beta, \alpha} \right) e_A{}^\alpha e_B{}^\beta . \tag{3.60}$$

We derive below Eq. (3.60) from Eq. (3.59). We begin by multiplying both sides of Eq. (3.59) by $e^C{}_\gamma$ and using Eq. (A.2). Thus

$$e_A{}^\alpha \, (\partial_\alpha \, e_B{}^\gamma) \, e^C{}_\gamma - e_B{}^\beta \, \left(\partial_\beta \, e_A{}^\gamma \right) e^C{}_\gamma = C^D{}_{AB} \, \delta_D{}^C = C^C{}_{AB} . \tag{3.61}$$

We also have the identities

$$\partial_\alpha \left(e_B{}^\gamma e^C{}_\gamma \right) = \partial_\alpha \delta_B{}^C = (\partial_\alpha e_B{}^\gamma) \, e^C{}_\gamma + e_B{}^\gamma \partial_\alpha e^C{}_\gamma = 0 , \tag{3.62}$$

$$\partial_\beta \left(e_A{}^\gamma e^C{}_\gamma \right) = \partial_\beta \delta_A{}^C = \left(\partial_\beta e_A{}^\gamma \right) e^C{}_\gamma + e_A{}^\gamma \partial_\beta e^C{}_\gamma = 0 . \tag{3.63}$$

Therefore Eq. (3.61) is

$$- e_A{}^\alpha e_B{}^\gamma e^C{}_{\gamma, \alpha} + e_B{}^\beta e_A{}^\gamma e^C{}_{\gamma, \beta} = C^C{}_{AB} , \tag{3.64}$$

which, after relabeling the dummy indices, reduces to Eq. (3.60).
Clearly

$$C^D{}_{AB} = - C^D{}_{BA} , \tag{3.65}$$

and we have,

$$C_{ABC} = \eta_{AD} C^D{}_{BC} , \tag{3.66}$$

$$C_{ABC} = -C_{ACB} . \tag{3.67}$$

Finally we give two expressions for the *Ricci rotation coefficients*, Γ_{ABC}. The Γ_{ABC} are related to the $\omega_{AB\mu}$, defined in Eq. (3.24), by the relation Eq. (3.68) below, where the metric signature affects the $\omega_{AB\mu}$ and hence the Γ_{ABC}.

$$\Gamma_{ABC} = \omega_{AB\mu} e_C{}^\mu . \tag{3.68}$$

Another expression for the Γ_{ABC}, which is written here *specifically for the metric signature* $(+, -, -, -)$ is

$$\Gamma_{ABC} = -\frac{1}{2} (C_{ABC} + C_{BCA} - C_{CAB}) . \tag{3.69}$$

Equations (3.68) and (3.69) agree with the definitions in Ref. [18], Eq. (253), p. 37, and Eq. (272), p. 39. For the metric signature $(-, +, +, +)$ one has to change the overall sign in Eq. (3.69), *in addition note* that the C's in Eq. (3.59) are the negatives of the C's defined in [19], Eq. (2.11). From Eq. (3.68) we see that

$$\Gamma_{ABC} = -\Gamma_{BAC} . \tag{3.70}$$

It is important to keep in mind the difference in index antisymmetry in Eqs. (3.67) and (3.70). Finally, using Eqs. (3.26), (3.28) and (3.68), we may now express the Fock-Ivanenko coefficients in terms of the Γ_{ABC},

$$\Gamma_C = \frac{\varepsilon}{4} \Gamma_{ABC} \gamma^A \gamma^B . \tag{3.71}$$

We have used as references, [16, 17, 19]. (Ref. [16] has some misprints in Sect. 11.4.) If one were to adopt the definitions and terminology of Soleng [19], one would have the advantage of being able to use the Mathematica package CARTAN (which is still available as of this writing) to calculate all of these quantities (symbolically) by computer.

Remark 8 Apart from being tetrad-dependent, it is clear from the above derivation, that the sign of the Ricci rotation coefficients, Γ_{ABC}, will depend on the metric signature since $C_{ABC} = \eta_{AD} C^D{}_{BC}$ (we add that certain authors, e.g., [17], define their spin connection Eq. (3.24) with the opposite sign). Furthermore, it is after Eq. (3.71), that is, when we write Eq. (3.30), that we have to choose a metric compatible representation of the γ matrices.

3.5   The Electromagnetic Interaction

As mentioned above, the requirement Eq. (3.20) is sufficient but not necessary. In general one may add a (possibly complex) [4] vector multiple of the unit matrix to the solution, Eq. (3.26). In this way we may generalize the Γ_μ's for the case where an arbitrary electromagnetic potential A_μ is present [4, 20]. We simply make the replacements:

$$\Gamma_\mu \rightarrow \Gamma_\mu + iq A_\mu I \,, \tag{3.72}$$

$$D_\mu \rightarrow D_\mu + iq A_\mu I \,, \tag{3.73}$$

where q is the charge of the particle described by ψ. Thus Eq. (3.32) is now generalized to

$$i\bar{\gamma}^\mu D_\mu \psi - m\psi = 0 \,, \tag{3.74}$$

$$i\bar{\gamma}^\mu \left(\partial_\mu + iq A_\mu + \Gamma_\mu \right) \psi - m\psi = 0 \,, \tag{3.75}$$

$$i\gamma^C \left(e_C + \frac{1}{4} \Gamma_{ABC} \gamma^A \gamma^B + iq A_C \right) \psi - m\psi = 0 \,, \tag{3.76}$$

where

$$\bar{\gamma}^\mu A_\mu = e_C{}^\mu \gamma^C A_\mu = \gamma^C e_C{}^\mu A_\mu = \gamma^C A_C \,. \tag{3.77}$$

This is consistent with the so-called *minimal coupling* procedure. One can easily deduce the correctness of this term by considering the Minkowski limit (e.g., see [21], pp. 64–67). For example, in the case of the hydrogenic atom, $q = -e, e > 0$, and in the standard notation the components A_C of electromagnetic potential due to the proton are

$$A = (A_0, A_1, A_2, A_3) = \left(\frac{Ze}{r}, 0, 0, 0 \right) \,, \tag{3.78}$$

so that $iq A_0 = -i \dfrac{Ze^2}{r}$.

Of course in curved spacetime one has to use appropriate Maxwell's equations. We refer the reader again to [4], Sect. VII, or [17].

3.6   The Newman-Penrose Formalism

In this section we give a short introduction to the Newman-Penrose formalism [22, 23]. Apart from the Newman-Penrose paper, we have found useful the exposition in the following texts [24–26]. However one has to be careful because, as usual, there are differences in some definitions and conventions among these references.

In the Newman-Penrose formalism the calculations are done using a complex null tetrad. One straightforward way to construct a complex null tetrad for a given metric, is to choose a set, e_A, of orthonormal tetrad vector fields (as discussed in Appendix A). These satisfy

$$\eta_{AB} = e_A{}^\alpha e_B{}^\beta g_{\alpha\beta} \,. \tag{3.79}$$

Then we define the complex null tetrad $l, n, m, \overline{m}$, below [27]:

$$\lambda_1 = l = \frac{1}{\sqrt{2}} (e_0 + e_3) \,, \tag{3.80}$$

$$\lambda_2 = n = \frac{1}{\sqrt{2}} (e_0 - e_3) \,, \tag{3.81}$$

$$\lambda_3 = m = \frac{1}{\sqrt{2}} (e_1 + ie_2) \,, \tag{3.82}$$

$$\lambda_4 = \overline{m} = \frac{1}{\sqrt{2}} (e_1 - ie_2) \,. \tag{3.83}$$

Note that $\overline{m} = m^*$, so that, in general, we will use A^* for the complex conjugate of A. Using Eq. (3.87) we can show that the null tetrad vectors $l, n, m, \overline{m}$, of Eqs. (3.80)–(3.83), satisfy the relations

$$l^\mu l_\mu = n^\mu n_\mu = \overline{m}^\mu \overline{m}_\mu = m^\mu m_\mu = 0 \,, \tag{3.84}$$

$$l^\mu m_\mu = n^\mu m_\mu = l^\mu \overline{m}_\mu = n^\mu \overline{m}_\mu = 0 \,, \tag{3.85}$$

$$l^\mu n_\mu = +1 \,, \quad m^\mu \overline{m}_\mu = -1 \,. \tag{3.86}$$

The frame field metric components, ζ_{AB}, for the λ_A of Eqs. (3.80)–(3.83), are given by

$$\zeta_{AB} = \lambda_A{}^\alpha \lambda_B{}^\beta g_{\alpha\beta} = \lambda_A{}^\alpha \lambda_{B\alpha} \,, \tag{3.87}$$

where $A, B = 1, 2, 3, 4$. We find that

$$\left(\zeta^{AB} \right) = \left(\zeta_{AB} \right) = \begin{pmatrix} 0 & 1 & 0 & 0 \\ 1 & 0 & 0 & 0 \\ 0 & 0 & 0 & -1 \\ 0 & 0 & -1 & 0 \end{pmatrix} . \tag{3.88}$$

Exercise 3 *Use Eq. (3.87) to derive Eq. (3.88). (Hint: Use Eqs. (3.80)–(3.83).)*

We also have the relations

$$g^{\alpha\beta} = \zeta^{AB} \lambda_A{}^\alpha \lambda_B{}^\beta \,, \tag{3.89}$$

$$\zeta^{CB} \zeta_{AB} = \lambda_A{}^\alpha \zeta^{CB} \lambda_{B\alpha} = \lambda_A{}^\alpha \lambda^C{}_\alpha = \delta_A{}^C \,. \tag{3.90}$$

In order to conform to the notation in the NP formalism literature, we shall change slightly the notation for the Ricci rotation coefficients given by Eq. (3.69), and write

$\Gamma_{ABC} = \gamma_{ABC}$ (with $A, B, C = 1, 2, 3, 4$ where 4 is the time label). The collection of equations below is very important and handy.

$$\lambda^A{}_\alpha = g_{\alpha\beta}\, \zeta^{AB} \lambda_B{}^\beta\,, \tag{3.91}$$

$$\gamma^D{}_{BC} = \lambda_B{}^\beta \lambda_C{}^\alpha \, \nabla_\alpha \, \lambda^D{}_\beta\,, \tag{3.92}$$

$$\gamma_{ABC} = \zeta_{AD}\, \gamma^D{}_{BC}\,, \tag{3.93}$$

and the equivalents to Eqs. (3.60), (3.65), and (3.69),

$$C^D{}_{BC} = \left(\lambda^D{}_{\alpha,\,\beta} - \lambda^D{}_{\beta,\,\alpha}\right) \lambda_B{}^\alpha \lambda_C{}^\beta\,, \tag{3.94}$$

$$C_{ABC} = \zeta_{AD}\, C^D{}_{BC}\,, \tag{3.95}$$

$$\gamma_{ABC} = -\frac{1}{2}\left(C_{ABC} + C_{BCA} - C_{CAB}\right)\,. \tag{3.96}$$

Remark 9 It is clear from Eqs. (3.80)–(3.83) that if the tetrad, e_A, is a Fermi tetrad, then the null tetrad, λ_B, is parallelly transported along the chosen congruence of timelike geodesics.

Remark 10 A choice of a null tetrad, λ_B, is equivalent to a choice of an orthonormal tetrad, e_A, since we can solve Eqs. (3.80)–(3.83), to express the e_A in terms of the λ_B, [27], thus,

$$e_0 = \frac{1}{\sqrt{2}}(l + n)\,, \tag{3.97}$$

$$e_1 = \frac{1}{\sqrt{2}}(m + \overline{m})\,, \tag{3.98}$$

$$e_2 = \frac{-i}{\sqrt{2}}(m - \overline{m})\,, \tag{3.99}$$

$$e_3 = \frac{1}{\sqrt{2}}(l - n)\,. \tag{3.100}$$

Remark 11 From Eq. (3.97) we may deduce the properties of the observer frame tetrad by finding the acceleration,

$$a = \nabla_{e_0} e_0\,. \tag{3.101}$$

Moreover, evaluating, $\nabla_{e_0} e_A$, $A = 0, 1, 2, 3$, will tell us whether the tetrad field is a Fermi tetrad or not (see Eq. (A.15)).

Exercise 4 *Starting with Eq. (3.89), show that*

$$g_{\alpha\beta} = n_\alpha l_\beta + l_\alpha n_\beta - \overline{m}_\alpha m_\beta - m_\alpha \overline{m}_\beta\,. \tag{3.102}$$

The null tetrad $l, n, m, \overline{m}$, is not uniquely defined by Eqs. (3.80)–(3.83). Without changing the direction of the field l, we may rescale it by an arbitrary factor A, where A is a non-vanishing real function. Thus

$$l'^\alpha = A l^\alpha. \tag{3.103}$$

This amounts to a reparametrization of the curves tangent to l. The vectors m and $\overline{m}$ may be rotated in their plane by an arbitrary angle ϕ; moreover, their scalar products with l, do not change when a multiple of l is added to them. Thus $m, \overline{m}$, are defined up to the transformations

$$m'^\alpha = e^{i\phi} m^\alpha + B l^\alpha, \tag{3.104}$$

where ϕ is a real function and B is a complex function. The remaining vector n, may be changed by a fixed multiple of l and a fixed multiple of a fixed vector in the $m, \overline{m}$ plane, so finally, we have [25]

$$n'^\alpha = \frac{1}{A} \left(n^\alpha + B^* e^{i\phi} m^\alpha + B e^{-i\phi} \overline{m}^\alpha + B B^* l^\alpha \right). \tag{3.105}$$

We can express these transformations as three classes of transformations [28],

$$l' = l, \; m' = m + Bl, \; n' = n + B^*m + B\overline{m} + BB^*l, \quad \text{(null rotation)}, \tag{3.106}$$
$$n' = n, \; m' = m + Bl, \; l' = l + B^*m + B\overline{m} + BB^*n, \quad \text{(null rotation)}, \tag{3.107}$$
$$l' = Al, \; m' = e^{i\phi}m, \; n' = A^{-1}n, \quad \text{(boost and orthogonal rotation)}. \tag{3.108}$$

We shall use the standard notation below, to designate the null tetrad vectors as directional derivatives

$$\lambda_1 = l = D, \quad \lambda_2 = n = \Delta, \quad \lambda_3 = m = \delta, \quad \lambda_4 = \overline{m} = \delta^*. \tag{3.109}$$

The above is expressed very clearly in reference [27]. "The role of the (vector) covariant derivative operator ∇_α is taken over in the NP formalism by four scalar operators:"

$$D = l^\alpha \nabla_\alpha, \quad \Delta = n^\alpha \nabla_\alpha, \quad \delta = m^\alpha \nabla_\alpha, \quad \delta^* = \overline{m}^\alpha \nabla_\alpha. \tag{3.110}$$

Thus, e.g., for any scalar function f we write

$$Df = l^\alpha f_{,\alpha}, \quad \Delta f = n^\alpha f_{,\alpha}, \quad \delta f = m^\alpha f_{,\alpha}, \quad \delta^* f = \overline{m}^\alpha f_{,\alpha}. \tag{3.111}$$

We give below the twelve so-called *spin coefficients* in terms of the Ricci rotation coefficients.

$$\kappa = \gamma_{311}, \quad \rho = \gamma_{314}, \quad \varepsilon = \tfrac{1}{2}\left(\gamma_{211} + \gamma_{341}\right),$$
$$\sigma = \gamma_{313}, \quad \mu = \gamma_{243}, \quad \gamma = \tfrac{1}{2}\left(\gamma_{212} + \gamma_{342}\right), \qquad (3.112)$$
$$\lambda = \gamma_{244}, \quad \tau = \gamma_{312}, \quad \alpha = \tfrac{1}{2}\left(\gamma_{214} + \gamma_{344}\right),$$
$$\nu = \gamma_{242}, \quad \pi = \gamma_{241}, \quad \beta = \tfrac{1}{2}\left(\gamma_{213} + \gamma_{343}\right).$$

Remark 12 One could choose the functions, A, B, ϕ, in Eqs. (3.103), (3.104), and (3.105), so as to maximize the number of vanishing the spin coefficients, thus possibly making the Dirac equation easier to solve. However it may not be easy to determine what observer frame the new tetrad corresponds to (see Remark 10).

We finally write the Dirac equation in the N-P formalism [18, 29–31].

$$(\Delta + \mu^* - \gamma^*)G_1 - (\delta^* + \beta^* - \tau^*)G_2 = i\mu_* F_1,$$
$$(D + \varepsilon^* - \rho^*)G_2 - (\delta + \pi^* - \alpha^*)G_1 = i\mu_* F_2, \qquad (3.113)$$
$$(D + \varepsilon - \rho)F_1 + (\delta^* + \pi - \alpha)F_2 = i\mu_* G_1,$$
$$(\Delta + \mu - \gamma)F_2 + (\delta + \beta - \tau)F_1 = i\mu_* G_2,$$

where $\mu_* \sqrt{2} = m =$ the mass of the particle. Equations (3.113) are in the chiral representation given by Eq. (D.18), see Ref. [32].

References

1. M.J. Duff, L.B. Okun, G. Veneziano, Trialogue on the number of fundamental constants. Inst. of Phys. Publishing 1–30, 9 March 2002, http://jhep.sissa.it/archive/papers/jhep032002023.pdf
2. E.A. Lord, *Tensors, Relativity and Cosmology* (Tata McGraw-Hill Publishing Co., Ltd., New Delhi, 1976)
3. B.F. Schutz, *A First Course in General Relativity*, 2nd edn. (Cambridge U. Press, New York, 2009)
4. L. Parker, One-electron atom as a probe of spacetime curvature. Phys. Rev. D **22**, 1922–1934 (1980)
5. T.C. Chapman, D.J. Leiter, On the generally covariant Dirac equation. Am. J. Phys. **44**, 858–862 (1976)
6. N.D. Birrell, P.C.W. Davies, *Quantum Fields in Curved Space* (Cambridge U. Press, New York, 1984)
7. S.A. Fulling, *Aspects of Quantum Field Theory in Curved Space-Time*, (Cambridge U. Press, New York, 1989), p. 167
8. R.A. Bertlmann, *Anomalies in Quantum Field Theory*, (Clarendon Press, Oxford, 1996), pp. 461–479
9. L. Parker, D. Toms *Quantum Field Theory in Curved Spacetime*, (Cambridge U. Press, Cambridge, 2009)
10. S. Carroll, *Spacetime and Geometry, an Introduction to General Relativity*, (Addison-Wesley, San Francisco, 2004)
11. D. Fursaev, D. Vassilevich, *Operators, Geometry and Quanta, Methods of Spectral Geometry in Quantum Field Theory*, (Springer, New York, 2011)

12. B.S. Kay, Linear spin-zero quantum fields in external gravitational and scalar fields. Commun. Math. Phys. **62**, 55–70 (1978)
13. E. Poisson, *A Relativist's Toolkit*, (Cambridge U. Press, Cambridge, 2004)
14. X. Huang, L. Parker, Hermiticity of the Dirac Hamiltonian in curved spacetime. Phys. Rev. D **79**, 024020 (2009)
15. P. Collas, D. Klein, Dirac particles in a gravitational shock wave. Class Quantum Grav. **35**, 125006 (2018), https://doi.org/10.1088/1361-6382/aac144
16. L. Ryder, *Introduction to General Relativity*, (Cambridge U. Press, Cambridge, 2009)
17. L.D. Landau, E.M. Lifshitz, *The Classical Theory of Fields*, (Butterworth-Heinemann, Oxford, 2003)
18. S. Chandrasekhar, *The Mathematical Theory of Black Holes*, (Clarendon Press, Oxford, 1992)
19. H.H. Soleng, *Tensors in Physics: the Mathematica package Cartan version 1.8* (Ad Infinitum AS, Fetsund, Norway, 2011). The Mathematica package is available at https://store.wolfram.com/view/book/D0709.str
20. M.K.-H. Kiessling, A.S. Tahvildar-Zadeh, The Dirac point electron in zero-gravity Kerr-Newman spacetime. J. Math. Phys. **56**, 042303 (2015)
21. C. Itzykson, J.-B. Zuber, *Quantum Field Theory*, (McGraw-Hill Inc., New York, 1980)
22. E. Newman, R. Penrose, An Approach to Gravitational Radiation by a Method of Spin Coefficients. J. Math. Phys. **3**, 566–578 (1962)
23. E. Newman, R. Penrose, Errata: an approach to gravitational radiation by a method of spin coefficients. J. Math. Phys. **4**, 998 (1963)
24. B. O'Neill, *The Geometry of Kerr Black Holes*, (A. K. Peters, Wellesley, 1995)
25. J. Plebański, A. Krasiński, *An Introduction to General Relativity and Cosmology*, (Cambridge U. Press, Cambridge, 2006)
26. E. Newman, R. Penrose, Spin-coefficient formalism. Scholarpedia **4**(6), 7445 (2009), http://www.scholarpedia.org/article/Spin-coefficient_formalism
27. R. Geroch, A. Held, R. Penrose, A space-time calculus based on pairs of null directions. J. Math. Phys. **14**, 874–881 (1973)
28. D. Bini, A. Geralico, R.T. Jantzen, Frenet-Serret formalism for null world lines, pp. 1–20. 19 Aug 2014, arXiv:1408.4281v1
29. S. Chandrasekhar, The solution of Dirac's equation in Kerr geometry. Proc. R. Soc. Lond. A. **349**, 571–575 (1976)
30. E. Montaldi, A. Zecca, Neutrino wave equation in the Robertson-Walker geometry. Int. J. Theor. Phys. **33**, 1053–1062 (1994)
31. A. Zecca, The Dirac equation in the Robertson-Walker space-time. J. Math. Phys. **37**, 874–879 (1996)
32. C. Röken, The Massive Dirac Equation in Kerr Geometry: Separability in Eddington-Finkelstein-Type Coordinates and Asymptotics, pp. 1–23. 6 Dec 2015, arXiv:1506.08038v2

Chapter 4
Examples in (3 + 1) GR

4.1 Schwarzschild Spacetime F-I

Important early calculations in $(3 + 1)$ cosmological spacetimes were carried out by Audretsch, and Schäfer [1, 2]. Several solutions in a non-cosmologial context are [3–10]. In our first example we shall calculate the Fock-Ivanenko coefficients and the resulting Dirac equations for the Schwarzschild metric (with signature $(+2)$). The γ matrices are in the standard representation, however because of our signature choice, we have to multiply each of the matrices (B.9), with a factor of $+i$. Then Eq. (B.9) change sign and Eq. (B.3) holds with $\varepsilon = +1$.

$$\{\gamma^A, \gamma^B\} = 2\eta^{AB} I \,. \tag{4.1}$$

In Appendix E.1 we show how these calculations below can be done using the Mathematica software package CARTAN. The Schwarzschild metric is

$$ds^2 = -\left(1 - \frac{2M}{r}\right) dt^2 + \frac{dr^2}{\left(1 - \frac{2M}{r}\right)} + r^2(d\theta^2 + \sin^2\theta d\phi^2) \,, \tag{4.2}$$

and we choose the orthonormal 1-forms and corresponding vectors below which satisfy $g_{\alpha\beta} = \eta_{AB} e^A{}_\alpha e^B{}_\beta$, (see Appendix A.1).

$$e^0 = \left(1 - \frac{2M}{r}\right)^{\frac{1}{2}} dt \,, \qquad e_0 = \left(1 - \frac{2M}{r}\right)^{-\frac{1}{2}} \partial_t \,, \tag{4.3}$$

$$e^1 = \left(1 - \frac{2M}{r}\right)^{-\frac{1}{2}} dr \,, \qquad e_1 = \left(1 - \frac{2M}{r}\right)^{\frac{1}{2}} \partial_r \,, \tag{4.4}$$

$$e^2 = r\, d\theta \,, \qquad e_2 = \frac{1}{r} \partial_\theta \,, \tag{4.5}$$

© The Author(s), under exclusive license to Springer Nature Switzerland AG 2019
P. Collas and D. Klein, *The Dirac Equation in Curved Spacetime*,
SpringerBriefs in Physics, https://doi.org/10.1007/978-3-030-14825-6_4

$$e^3 = r \sin \theta \, d\phi, \qquad\qquad e_3 = \frac{1}{r \sin \theta} \partial_\phi. \qquad (4.6)$$

Equations (4.3)–(4.6) are the tetrad 1-form fields and vectors of an observer with $\dot{r} = 0$, $\dot{\theta} = 0$, $\dot{\phi} = 0$. Using Eq. (3.24) we find the nonvanishing spin connection coefficients,

$$\omega_{10t} = \frac{M}{r^2}, \qquad (4.7)$$

$$\omega_{21\theta} = \left(1 - \frac{2M}{r}\right)^{\frac{1}{2}}, \qquad (4.8)$$

$$\omega_{31\phi} = \sin \theta \left(1 - \frac{2M}{r}\right)^{\frac{1}{2}}, \qquad (4.9)$$

$$\omega_{32\phi} = \cos \theta. \qquad (4.10)$$

We now use Eq. (3.26) with $\varepsilon = +1$ and (3.28), to obtain the Fock-Ivanenko coefficients Γ_C.

$$\Gamma_0 = -\frac{M}{2r^2} \left(1 - \frac{2M}{r}\right)^{-\frac{1}{2}} \gamma^0 \gamma^1, \qquad (4.11)$$

$$\Gamma_1 = 0, \qquad (4.12)$$

$$\Gamma_2 = -\frac{1}{2r} \left(1 - \frac{2M}{r}\right)^{\frac{1}{2}} \gamma^1 \gamma^2, \qquad (4.13)$$

$$\Gamma_3 = -\frac{1}{2r} \left(1 - \frac{2M}{r}\right)^{\frac{1}{2}} \gamma^1 \gamma^3 - \frac{\cot \theta}{2r} \gamma^2 \gamma^3. \qquad (4.14)$$

The Dirac equation, *is not* Eq. (3.30), but rather the one below because the i factor has been absorbed in γ^A.

$$\begin{aligned}
\big[\gamma^0 \left(e_0{}^t \, \partial_t + \Gamma_0\right) + \gamma^1 e_1{}^r \, \partial_r + \gamma^2 \left(e_2{}^\theta \, \partial_\theta + \Gamma_2\right) \\
+ \gamma^3 \left(e_3{}^\phi \, \partial_\phi + \Gamma_3\right)\big]\psi - m\psi = 0.
\end{aligned} \qquad (4.15)$$

From Eq. (4.11) we see that the term $\gamma^0 \Gamma_0$ appearing in Eq. (4.15) simplifies to

$$\gamma^0 \Gamma_0 = \frac{M}{2r^2} \left(1 - \frac{2M}{r}\right)^{-\frac{1}{2}} \gamma^1, \qquad (4.16)$$

and so on.

4.2 Schwarzschild Spacetime N-P

In this subsection we compare the two approaches, N-P and F-I. In Appendix E.2 we show how the calculations below can be done using the Mathematica software package CARTAN. Here we use the metric signature $(+, -, -, -)$ and so we write the Schwarzschild metric,

$$ds^2 = \left(1 - \frac{2M}{r}\right) dt^2 - \frac{dr^2}{\left(1 - \frac{2M}{r}\right)} - r^2(d\theta^2 + \sin^2\theta d\phi^2). \qquad (4.17)$$

We adopt essentially Chandrasekhar's null tetrad cf., [11], Eq. (281) p. 134, (we differ only by a $\sqrt{2}$). We have $l = (l^t, l^r, l^\theta, l^\phi)$, etc.,

$$l = \frac{1}{\sqrt{2}} \left(\frac{1}{X}, 1, 0, 0\right), \qquad (4.18)$$

$$n = \frac{1}{\sqrt{2}} (1, -X, 0, 0), \qquad (4.19)$$

$$m = \frac{1}{\sqrt{2}} \left(0, 0, \frac{1}{r}, \frac{i}{r\sin\theta}\right), \qquad (4.20)$$

$$\overline{m} = \frac{1}{\sqrt{2}} \left(0, 0, \frac{1}{r}, \frac{-i}{r\sin\theta}\right), \qquad (4.21)$$

where

$$X = 1 - \frac{2M}{r}. \qquad (4.22)$$

We find the spin coefficients below.

$$\alpha = -\frac{\cot\theta}{2\sqrt{2}\,r}, \qquad (4.23)$$

$$\beta = \frac{\cot\theta}{2\sqrt{2}\,r}, \qquad (4.24)$$

$$\gamma = \frac{M}{\sqrt{2}\,r^2}, \qquad (4.25)$$

$$\mu = -\frac{X}{\sqrt{2}\,r}, \qquad (4.26)$$

$$\rho = -\frac{1}{\sqrt{2}\,r}, \qquad (4.27)$$

Now we use Eqs. (3.111) and (3.113) and write below the resulting four Dirac equations for those who wish to check their results.

$$\partial_t G_1 - X\partial_r G_1 + \frac{M-r}{r^2} G_1 - \frac{1}{r}\partial_\theta G_2 + \frac{i}{r\sin\theta}\partial_\phi G_2$$
$$- \frac{\cot\theta}{2r} G_2 = im F_1, \qquad (4.28)$$

$$\frac{1}{X}\partial_t G_2 + \partial_r G_2 + \frac{1}{r} G_2 - \frac{1}{r}\partial_\theta G_1 - \frac{i}{r\sin\theta}\partial_\phi G_1$$
$$- \frac{\cot\theta}{2r} G_1 = im F_2, \qquad (4.29)$$

$$\frac{1}{X}\partial_t F_1 + \partial_r F_1 + \frac{1}{r} F_1 + \frac{1}{r}\partial_\theta F_2 - \frac{i}{r\sin\theta}\partial_\phi F_2$$
$$+ \frac{\cot\theta}{2r} F_2 = im G_1, \qquad (4.30)$$

$$\partial_t F_2 - X\partial_r F_2 + \frac{M-r}{r^2} F_2 + \frac{1}{r}\partial_\theta F_1 + \frac{i}{r\sin\theta}\partial_\phi F_1$$
$$+ \frac{\cot\theta}{2r} F_1 = im G_2, \qquad (4.31)$$

where the Dirac wavefunction is

$$\psi = \begin{pmatrix} F_1 \\ F_2 \\ G_1 \\ G_2 \end{pmatrix}. \qquad (4.32)$$

To do the calculations following the Fock-Ivanenko approach, we recall Remark 10 and use Eqs. (3.97)–(3.100), to obtain the orthogonal tetrad corresponding to the null tetrad given by Eqs. (4.18)–(4.21). We have

$$e_0 = \left(\frac{1+X}{2X}, \frac{1-X}{2}, 0, 0 \right), \qquad (4.33)$$

$$e_1 = \left(0, 0, \frac{1}{r}, 0 \right), \qquad (4.34)$$

$$e_2 = \left(0, 0, 0, \frac{1}{r\sin\theta} \right), \qquad (4.35)$$

$$e_3 = \left(\frac{1-X}{2X}, \frac{1+X}{2}, 0, 0 \right), \qquad (4.36)$$

where $e_A = (e_A{}^t, e_A{}^r, e_A{}^\theta, e_A{}^\phi)$, and X is given by Eq. (4.22). Recalling Remark 11, we find that $a = \nabla_{e_0} e_0 \neq 0$, so this is not a freely falling particle frame. Finally the Dirac equations obtained with this approach are, of course, identical to Eqs. (4.28), (4.31).

4.3 Nonfactorizable Metric

In this example we shall write the Dirac equation in the spacetime considered by Hounkonnou and Mendy [12]

$$ds^2 = -dt^2 + a^2(t)\left(dx^2 + b^2(x)\left[dy^2 + c^2(y)dz^2\right]\right). \tag{4.37}$$

We remark that the de Sitter universe metric and the usual Friedman-Lemaître-Robertson-Walker metric of standard cosmology, for each curvature parameter k separately, are special cases of the above form.

As in the example of Sect. 4.1 we again choose the tetrad for an observer with $\dot{x} = 0$, $\dot{y} = 0$, $\dot{z} = 0$. The 1-forms and corresponding vectors are given below.

$$e^0 = dt\,, \qquad\qquad\qquad e_0 = \partial_t\,, \tag{4.38}$$

$$e^1 = a(t)\,dx\,, \qquad\qquad e_1 = \frac{1}{a(t)}\,\partial_x\,, \tag{4.39}$$

$$e^2 = a(t)b(x)\,dy\,, \qquad\quad e_2 = \frac{1}{a(t)b(x)}\,\partial_y\,, \tag{4.40}$$

$$e^3 = a(t)b(x)c(y)\,dz\,, \qquad e_3 = \frac{1}{a(t)b(x)c(y)}\,\partial_z\,. \tag{4.41}$$

Using Eq. (3.24) we find the nonvanishing spin connection coefficients,

$$\omega_{10x} = a_{,t}\,, \qquad \omega_{20y} = b\,a_{,t}\,, \qquad \omega_{21y} = b_{,x}\,, \tag{4.42}$$

$$\omega_{30z} = b\,c\,a_{,t}\,, \qquad \omega_{31z} = c\,b_{,x}\,, \qquad \omega_{32z} = c_{,z}\,. \tag{4.43}$$

Again we use Eqs. (3.26) and (3.28), to obtain the the Fock-Ivanenko coefficients Γ_C.

$$\Gamma_0 = 0\,, \tag{4.44}$$

$$\Gamma_1 = -\frac{a_{,t}}{2a}\,\gamma^0\gamma^1\,, \tag{4.45}$$

$$\Gamma_2 = -\frac{a_{,t}}{2a}\,\gamma^0\gamma^2 - \frac{b_{,x}}{2ab}\,\gamma^1\gamma^2\,, \tag{4.46}$$

$$\Gamma_3 = -\frac{a_{,t}}{2a}\,\gamma^0\gamma^3 - \frac{b_{,x}}{2ab}\,\gamma^1\gamma^3 - \frac{c_{,y}}{2abc}\,\gamma^2\gamma^3\,. \tag{4.47}$$

Note that Hounkonnou and Mendy in Ref. [12], define their Γ_μ with opposite sign from the one adopted here, Eq. (3.26), and in the first part of their paper they effectively multiply their γ matrices by $(-i)$. Using Eq. (C.11) and the relations (B.19), we write the resulting Dirac equation as,

$$\gamma^C (e_C + \Gamma_C) \psi + m\psi = \left[\gamma^0 \left(\partial_t + \frac{3a_{,t}}{2a} \right) + \gamma^1 \left(\frac{1}{a}\partial_x + \frac{b_{,x}}{ab} \right) \right.$$
$$\left. + \gamma^2 \left(\frac{1}{ab}\partial_y + \frac{c_{,y}}{2abc} \right) + \gamma^3 \frac{1}{abc} \partial_z \right] \psi + m\psi = 0.$$

(4.48)

A further simplification is achieved if we let

$$\psi \equiv a^{-\frac{3}{2}} b^{-1} c^{-\frac{1}{2}} \Psi. \tag{4.49}$$

A short calculation shows that we may now rewrite the Dirac equation in the simplified form

$$\left[\gamma^0 \partial_t + \frac{1}{a}\gamma^1 \partial_x + \frac{1}{ab}\gamma^2 \partial_y + \frac{1}{abc}\gamma^3 \partial_z \right] \Psi + m\Psi = 0. \tag{4.50}$$

At some point in Ref. [12] the authors specifically adopt the Jauch-Rohrlich representation of the γ matrices discussed in Sect. B.5. Further clarification may be obtained by reviewing Sect. B.3.

4.4   de Sitter Spacetime, Fermi Coordinates

Fermi coordinates in general relativity are the closest approximation to Cartesian coordinates in special relativity. Thus they are a generalization of the notion of an inertial frame of reference in Minkowski spacetime, to curved spacetime. The observer's frame (path) is a timelike geodesic, and the spacelike coordinates are orthogonal to this path. Fermi coordinates are usually admissible only in a finite cylindrical region, $\mathcal{C}$, about the observer's timelike geodesic. In cases where analytic solutions of sets of timelike and spacelike geodesics can be found, several exact coordinate transformations from the original coordinates to Fermi coordinates have been found, [13–20].

In this section we shall consider the Dirac equation in the de Sitter universe using *exact* (global) Fermi coordinates (x^0, x^1, x^2, x^3) *with respect to the reference observer* $(\tau, 0, 0, 0)$. We shall first write the de Sitter metric in the standard coordinates used in Ref. [15], (but we adopt a different notation from the one used there in order to avoid confusion).

$$ds^2 = -d(y^0)^2 + e^{2ay^0} \delta_{ij} dy^i dy^j, \tag{4.51}$$

A Fermi tetrad field on the set of geodesics $\gamma(\tau) = (\tau, \, y^1{}_0, \, y^2{}_0, \, y^3{}_0)$, i.e., $y^0 = \tau$ and $y^i = $ const., is

$$\lambda_0 = \partial_{y^0}, \quad \lambda_I = e^{-a\tau}\partial_{y^I}, \quad I = (1, 2, 3). \tag{4.52}$$

We transform the metric of Eq. (4.51) to the metric in global Fermi coordinates using the transformation derived in [15, 17], relating the y^μ to the Fermi x^μ,

$$e^{ay^0} = e^{ax^0}\cos{(a\rho)} \tag{4.53}$$

$$y^i = e^{-ax^0}\frac{\tan{(a\rho)}}{a\rho}x^i, \tag{4.54}$$

where $\rho = \sqrt{\delta_{ij}x^i x^j}$, $a = \sqrt{\Lambda/3}$, and $0 \leq \rho < \pi/(2a)$.

The resulting metric is,

$$ds^2 = -\cos^2{(a\rho)}\,d(x^0)^2 + \left[\frac{x^i x^j}{\rho^2} + \frac{\sin^2{(a\rho)}}{a^2\rho^2}\left(\delta_{ij} - \frac{x^i x^j}{\rho^2}\right)\right]dx^i dx^j, \tag{4.55}$$

We may obtain a set of Fermi tetrad 1-form field in Fermi coordinates by transforming the 1-forms corresponding to the vectors of Eq. (4.52), using the transformation Eqs. (4.53), (4.54). The Fermi tetrad field obtained is complicated because the set of geodesics corresponding to the set $\gamma(\tau)$ has lost its original simplicity in the x^μ coordinates. We can find the inverse of the transformation Eqs. (4.53), (4.54), which we shall refer to as F, thus

$$x^0 = \left(\frac{1}{a}\right)\ln{\left(\frac{e^{ay^0}}{\sqrt{1 - a^2 e^{2ay^0}R^2}}\right)}, \tag{4.56}$$

$$x^i = \frac{\arccos{\sqrt{1 - a^2 e^{2ay^0}R^2}}}{aR}y^i, \tag{4.57}$$

where $R = \sqrt{\delta_{ij}y^i y^j}$. We know that if $F : M \to N$ is an isometry and γ is a geodesic in M, then $F \circ \gamma$ is a geodesic in N. Therefore the set of geodesics, $\gamma(\tau)$, are now given by (recall that $y^0 = \tau$)

$$F \circ \gamma = \left(x^0(\tau), \, x^1(\tau), \, x^2(\tau), \, x^3(\tau)\right). \tag{4.58}$$

The above complications do not prevent us from carrying on with our calculations for the Dirac equation. We give the set of tetrad 1-forms we obtained below.

$$e^0 = dx^0 - f(\rho)\left(x^1\,dx^1 + x^2\,dx^2 + x^3\,dx^3\right), \tag{4.59}$$

$$e^1 = -x^1 h(\rho)\,dx^0$$

$$+ \frac{a\rho\left(x^1\right)^2 \sec(a\rho) + \left[\left(x^2\right)^2 + \left(x^3\right)^2\right]\sin(a\rho)}{a\rho^3}\,dx^1$$

$$+ x^1 x^2\, p(\rho)\,dx^2 + x^1 x^3\, p(\rho)\,dx^3, \tag{4.60}$$

$$e^2 = -x^2 h(\rho)\,dx^0 + x^1 x^2\, p(\rho)\,dx^1$$

$$+ \frac{a\rho\left(x^2\right)^2 \sec(a\rho) + \left[\left(x^1\right)^2 + \left(x^3\right)^2\right]\sin(a\rho)}{a\rho^3}\,dx^2$$

$$+ x^2 x^3\, p(\rho)\,dx^3, \tag{4.61}$$

$$e^3 = -x^3 h(\rho)\,dx^0 + x^1 x^3\, p(\rho)\,dx^1 + x^2 x^3\, p(\rho)\,dx^2$$

$$+ \frac{a\rho\left(x^3\right)^2 \sec(a\rho) + \left[\left(x^1\right)^2 + \left(x^2\right)^2\right]\sin(a\rho)}{a\rho^3}\,dx^3, \tag{4.62}$$

where

$$f(\rho) = \frac{\tan(a\rho)}{\rho}, \tag{4.63}$$

$$h(\rho) = \frac{\sin(a\rho)}{\rho}, \tag{4.64}$$

$$p(\rho) = \left(\frac{\sec(a\rho)}{\rho^2} - \frac{\sin(a\rho)}{a\rho^3}\right), \tag{4.65}$$

$$q(\rho) = \left(-\frac{a\csc(a\rho)}{\rho} + \frac{\sec(a\rho)}{\rho^2}\right). \tag{4.66}$$

The corresponding tetrad vectors are,

$$e_0 = \sec^2(a\rho)\,\partial_{x^0} + f(\rho)\left(x^1\,\partial_{x^1} + x^2\,\partial_{x^2} + x^3\,\partial_{x^3}\right), \tag{4.67}$$

$$e_1 = x^1 f(\rho)\sec(a\rho)\,\partial_{x^0}$$

$$+ \frac{a\rho\left[\left(x^2\right)^2 + \left(x^3\right)^2\right]\csc(a\rho) + \left(x^1\right)^2 \sec(a\rho)}{\rho^2}\,\partial_{x^1}$$

$$+ q(\rho)\left(x^1 x^2\,\partial_{x^2} + x^1 x^3\,\partial_{x^3}\right), \tag{4.68}$$

$$e_2 = x^2 f(\rho)\sec(a\rho)\,\partial_{x^0} + x^1 x^2\, q(\rho)\,\partial_{x^1}$$

$$+ \frac{a\rho\left[\left(x^1\right)^2 + \left(x^3\right)^2\right]\csc(a\rho) + \left(x^2\right)^2 \sec(a\rho)}{\rho^2}\,\partial_{x^2}$$

$$+ x^2 x^3 q(\rho) \partial_{x^3}, \tag{4.69}$$

$$e_3 = x^3 f(\rho) \sec(a\rho) \partial_{x^0} + q(\rho) \left(x^1 x^3 \partial_{x^1} + x^2 x^3 \partial_{x^2} \right)$$

$$+ \frac{a\rho \left[\left(x^1 \right)^2 + \left(x^2 \right)^2 \right] \csc(a\rho) + \left(x^3 \right)^2 \sec(a\rho)}{\rho^2} \partial_{x^3}. \tag{4.70}$$

We now obtain the nonvanishing Ricci rotation coefficients, Γ_{ABC}, of Eqs. (3.68), (3.69), (but see comment below (3.69). Then the Fock-Ivanenko coefficients, Γ_C, of Eq. (3.71). We have

$$\Gamma_{101} = \Gamma_{202} = \Gamma_{303} = a, \tag{4.71}$$

and (recall Proposition 2) we find that

$$\Gamma_0 = 0, \quad \Gamma_A = -\frac{a}{2} \gamma^0 \gamma^A, \quad A = 1, 2, 3. \tag{4.72}$$

Using Eq. (4.72), we write the Dirac equation (C.10),

$$\left[\gamma^0 e_0 + \gamma^1 \left(e_1 - \frac{a}{2} \gamma^0 \gamma^1 \right) + \gamma^2 \left(e_2 - \frac{a}{2} \gamma^0 \gamma^2 \right) \right.$$

$$\left. + \gamma^3 \left(e_3 - \frac{a}{2} \gamma^0 \gamma^3 \right) \right] \psi - m\psi = 0. \tag{4.73}$$

We note that because of our $(-, +, +, +,)$ signature, the γ matrices will satisfy Eq. (B.19) and the Dirac equation, simplifies to

$$\left[\gamma^A e_A + \gamma^0 \left(e_0 + \frac{3a}{2} \right) \right] \psi - m\psi = 0, \quad A = 1, 2, 3. \tag{4.74}$$

Equation (4.74) looks deceptively simple, but the complications arise from the expressions for the e_A, therefore it may be better to work in the original coordinates, Eq. (4.51) with the original tetrad Eq. (4.52).

References

1. J. Audretsch, G. Schäfer, Quantum mechanics of electromagnetically bounded spin-$\frac{1}{2}$ particles in an expanding universe: I. Influence of the expansion. Gen. Relativ. Gravit. **9**, 243–255 (1978)
2. J. Audretsch, G. Schäfer, Quantum mechanics of electromagnetically bounded spin-$\frac{1}{2}$ particles in expanding universes: II. Energy spectrum of the hydrogen atom. Gen. Relativ. Gravit. **9**, 489–500 (1978)
3. S. Chandrasekhar, The solution of Dirac's equation in Kerr geometry. Proc. R. Soc. Lond. A. **349**, 571–575 (1976)
4. D.N. Page, Dirac equation around a charged, rotating black hole. Phys. Rev. D **14**, 1509–1510 (1976)

5. V.M. Villalba, Exact solution to Dirac equation in a reducible Einstein space. J. Math Phys. **31**, 1483–1486 (1990)
6. D.-Y. Chen, Q.-Q. Jiang, X.-T. Zu, Hawking radiation of Dirac particles via tunnelling from rotating black holes in de Sitter spaces. Phys. Lett. B **665**, 106–110 (2008)
7. H. Cebeci, N. Özdemir, Dirac equation in Kerr-Taub NUT spacetime. Class. Quantum Grav. **30**, 175005 (2013)
8. O.V. Veko, E.M. Ovsiyuk, V.M. Red'kov, Dirac particle in the presence of a magnetic charge in de Sitter universe: exact solutions and transparency of the cosmological horizon. Nonlinear Phenom. Complex Syst. **17**, 461–463 (2014)
9. L. Anderson, C. Bär, Wave and Dirac equations manifolds, pp. 1–21. 17 April 2018, arXiv:1710.04512v2
10. J.F. García, C. Sabín, Dirac equation in exotic space-times, pp. 1–7. 11 Nov 2018, arXiv:1811.00385v2
11. S. Chandrasekhar, *The Mathematical Theory of Black Holes*, (Clarendon Press, Oxford, 1992)
12. M.N. Hounkonnou, J.E.B. Mendy, Exact solutions of the Dirac equation in a nonfactorizable metric. J. Math. Phys. **40**, 3827–3842 (1999)
13. L. Parker, One-electron atom as a probe of spacetime curvature. Phys. Rev. D **22**, 1922–1934 (1980)
14. P. Collas, D. Klein, Dirac particles in a gravitational shock wave. Class. Quantum Grav. **35**, 125006 (2018), https://doi.org/10.1088/1361-6382/aac144
15. C. Chicone, B. Mashhoon, Explicit Fermi coordinates and tidal dynamics in de Sitter and Gödel spacetimes. Phys. Rev. D **74**, 064019 (2006)
16. D. Klein, P. Collas, General transformation formulas for Fermi-Walker coordinates. Class. Quantum Grav. **25**, 145019 (2008), arXiv:0712.3838
17. D. Klein, P. Collas, Exact Fermi coordinates for a class of space-times. J. Math. Phys. **51**, 022501 (2010)
18. D. Klein, E. Randles, Fermi coordinates, simultaneity, and expanding space in Robertson-Walker cosmologies. Ann. Henri Poincaré **12**, 303–328 (2011)
19. D. Bini, A. Geralico, R.T. Jantzen, Fermi coordinates in Schwarzschild spacetime: closed form expressions. Gen. Relativ. Gravit. **43**, 1837–1853 (2011)
20. D. Klein, Maximal Fermi charts and geometry of inflationary universes. Ann Henri Poincaré (2012). https://doi.org/10.1007/s00023-012-0227-3

Chapter 5
The Dirac Equation in $(1 + 1)$ GR

5.1 Introduction to $(1 + 1)$

The importance of working in $(1 + 1)$ gravity for modeling its quantum versions, was originally pointed out by Teitelboim [1] and Jackiw [2]. We recall that the only non-trivial version of Einstein's equations is

$$R - \Lambda = 8\pi T, \tag{5.1}$$

where are R is the scalar curvature and Λ the cosmological constant and T is the trace of the $(1 + 1)$ stress-energy tensor. When source terms are added to Eq. (5.1), several nontrivial solutions are obtained including black hole spacetimes [3–7]. It is useful to recall the well-known result from differential geometry, namely, that all two-dimensional manifolds are conformally flat, therefore one may always choose coordinates so that the metric is

$$g_{\alpha\beta} = e^{\varphi}\eta_{\alpha\beta}. \tag{5.2}$$

Then Eq. (5.1) reduces to the flat space Liouville equation,

$$\eta^{\alpha\beta}\partial_\alpha\partial_\beta\varphi + \Lambda e^{\varphi} = 0, \tag{5.3}$$

which has to be solved for the conformal factor φ.

The interested reader should read the excellent presentation in the relevant sections of Brown's textbook, *Lower Dimensional Gravity* [8]. Other relevant applications and reviews of the subject can be found in [9–11]. In addition we give some references to papers dealing specifically with the Dirac equation in $(1 + 1)$ curved spacetime, [12–18].

In this section we shall adopt the metric signature $(+, -)$. In $(1 + 1)$ general relativity the Dirac equation simplifies and may be written as follows [12–14].

© The Author(s), under exclusive license to Springer Nature Switzerland AG 2019
P. Collas and D. Klein, *The Dirac Equation in Curved Spacetime*,
SpringerBriefs in Physics, https://doi.org/10.1007/978-3-030-14825-6_5

$$\left[i\gamma^A e_A{}^\mu \partial_\mu + \frac{i}{2}\gamma^A \frac{1}{\sqrt{-g}}\partial_\mu \left(\sqrt{-g}\, e_A{}^\mu\right) - m I_2\right]\psi = 0, \tag{5.4}$$

where the zweibein vector label A runs over 0, 1, and for the spinor we write

$$\psi = \begin{pmatrix} \psi_1 \\ \psi_2 \end{pmatrix}. \tag{5.5}$$

In what follows we will further restrict ourselves to the *chiral* (Weyl) representation of the Dirac γ matrices, specifically we choose [12]

$$\gamma^0 = \begin{pmatrix} 0 & 1 \\ 1 & 0 \end{pmatrix}, \quad \gamma^1 = \begin{pmatrix} 0 & -1 \\ 1 & 0 \end{pmatrix}. \tag{5.6}$$

Thus

$$\left(\gamma^0\right)^2 = I_2, \quad \left(\gamma^1\right)^2 = -I_2, \tag{5.7}$$

and

$$\{\gamma^A, \gamma^B\} = 2\eta^{AB} I_2. \tag{5.8}$$

We also define the *chirality operator* [19],

$$\gamma^5 := \gamma^0 \gamma^1 = \begin{pmatrix} 1 & 0 \\ 0 & -1 \end{pmatrix}. \tag{5.9}$$

One advantage of the chiral representation is the ease of decoupling of Eq. (5.4). Of course the spinor wave function components of Eq. (5.5) are now eigenstates of the chirality operator γ^5, so we may write

$$\psi = \begin{pmatrix} \psi_{(+)} \\ \psi_{(-)} \end{pmatrix}, \tag{5.10}$$

with the eigenvalues $\gamma^5 \psi_{(+)} = +\psi_{(+)}$, and, $\gamma^5 \psi_{(-)} = -\psi_{(-)}$.

5.2   The Dirac Equation in the Milne Universe

We shall consider solutions of the Dirac equation in the Milne universe in two different charts: (a) in standard comoving coordinates (t, x), in which case the metric is

$$ds^2 = dt^2 - a_0^2 t^2 dx^2, \tag{5.11}$$

and (b) in exact Fermi coordinates (τ, ρ), in which case the Milne universe is the interior of the forward lightcone of Minkowski spacetime, [20, 21] thus

$$ds^2 = d\tau^2 - d\rho^2, \tag{5.12}$$

where $\tau > |\rho|$.

(a) We use the default zweibein

$$\bar{e}_0 = \partial_t, \qquad \bar{e}_1 = \frac{1}{a_0 t} \partial_x. \tag{5.13}$$

Since the metric in Eq. (5.11) does not depend on x the corresponding canonical momentum p_x is a constant both in classical and quantum mechanics. We take advantage of this fact and write the 2-component spinor ψ as

$$\psi(t, x) = \begin{pmatrix} \psi_1 \\ \psi_2 \end{pmatrix} = e^{-ip_x x} \begin{pmatrix} f_1(t) \\ f_2(t) \end{pmatrix}, \tag{5.14}$$

where p_x is the the particle's momentum. So the spinor $\psi(t, x)$ is an eigenstate of the chirality operator γ^5 and an eigenstate of p_x.

Remark 13 Note that neither $p_t = p^t$ nor $p^x = -p_x/(a_0^2 t^2)$ are conserved.

One finds that the only nonvanishing term from the second set of terms in Eq. (5.4) is

$$\frac{i}{2}\gamma^0 \frac{1}{a_0 t} \left[\partial_t \left(a_0 t\, \bar{e}_0{}^t \right) \right] = \frac{i}{2t}\gamma^0. \tag{5.15}$$

Thus Eq. (5.4) reduces to

$$\left[i\gamma^0 \partial_t + \frac{i}{a_0 t} \gamma^1 \partial_x + \frac{i}{2t}\gamma^0 - m I_2 \right] \psi = 0. \tag{5.16}$$

Now we substitute Eqs. (5.6) and (5.14) in Eq. (5.16) and obtain the coupled equations,

$$f_1 = \frac{1}{m} \left(i\partial_t - \frac{p_x}{a_0 t} + \frac{i}{2t} \right) f_2, \tag{5.17}$$

$$f_2 = \frac{1}{m} \left(i\partial_t + \frac{p_x}{a_0 t} + \frac{i}{2t} \right) f_1. \tag{5.18}$$

Finally, decoupling Eqs. (5.17) and (5.18), we obtain

$$t^2 f_1'' + t f_1' + \left[m^2 t^2 - \left(\frac{1}{2} - \frac{i p_x}{a_0} \right)^2 \right] f_1 = 0, \tag{5.19}$$

where the primes denote derivatives with respect to t. The solution of this equation is given below in terms of the Bessel functions J_ν and Y_ν of the first and second kind respectively,

$$f_1(t) = A J_\nu(mt) + B Y_\nu(mt), \tag{5.20}$$

where A and B are arbitrary constants and

$$\nu = \frac{1}{2} - \frac{i p_x}{a_0}. \tag{5.21}$$

Using Eq. (5.18), we find that

$$f_2(t) = i A J_{\nu-1}(mt) + i B Y_{\nu-1}(mt). \tag{5.22}$$

(b) Now we transform the solution to the exact Fermi coordinates (τ, ρ). The transformation and its inverse is given by [20]

$$t = \sqrt{\tau^2 - \rho^2}, \tag{5.23}$$

$$x = \left(\frac{1}{a_0}\right) \tanh^{-1}\left(\frac{\rho}{\tau}\right), \tag{5.24}$$

$$\tau = t \cosh(a_0 x), \tag{5.25}$$

$$\rho = t \sinh(a_0 x). \tag{5.26}$$

Under the above coordinate transformation the original zweibein 1-form fields,

$$\bar{e}^0 = dt, \qquad \bar{e}^1 = a_0 t \, dx, \tag{5.27}$$

transform into the 1-form fields h^A below

$$h^0 = \frac{\tau}{\sqrt{\tau^2 - \rho^2}} d\tau - \frac{\rho}{\sqrt{\tau^2 - \rho^2}} d\rho, \tag{5.28}$$

$$h^1 = \frac{-\rho}{\sqrt{\tau^2 - \rho^2}} d\tau + \frac{\tau}{\sqrt{\tau^2 - \rho^2}} d\rho, \tag{5.29}$$

which, of course, satisfy the relation

$$h^A{}_\alpha h^B{}_\beta \eta^{\alpha\beta} = \eta^{AB}, \tag{5.30}$$

where the upper case latin indices run over 0, 1, while the greek indices run over τ, ρ. We shall write $\psi_h(\tau, \rho)$ for the $\psi(t, x)$ of Eq. (5.14) transformed using Eqs. (5.23) and (5.24). Thus

$$\psi_h(\tau, \rho) = e^{-i p_x \left(\frac{1}{a_0}\right) \tanh^{-1}\left(\frac{\rho}{\tau}\right)} \begin{pmatrix} f_1(\sqrt{\tau^2 - \rho^2}) \\ f_2(\sqrt{\tau^2 - \rho^2}) \end{pmatrix}. \tag{5.31}$$

We will now perform a local Lorentz transformation, Λ, which will transform the zweibein 1-form fields h^A into the canonical zweibein for the metric (5.12),

$$e^0 = d\tau, \qquad e^1 = d\rho. \tag{5.32}$$

The transformation Λ is given by

$$e^A{}_\alpha = \Lambda^A{}_B\, h^B{}_\alpha. \tag{5.33}$$

We find

$$\Lambda = \begin{pmatrix} \Lambda^0{}_0 & \Lambda^0{}_1 \\ \Lambda^1{}_0 & \Lambda^1{}_1 \end{pmatrix} = \frac{1}{\sqrt{\tau^2 - \rho^2}} \begin{pmatrix} \tau & \rho \\ \rho & \tau \end{pmatrix}. \tag{5.34}$$

We can calculate the matrix L used in Eq. 3.6 following the prescription given in Appendix B of [22],

$$L = \begin{pmatrix} \left(\frac{\tau+\rho}{\tau-\rho}\right)^{1/4} & 0 \\ 0 & \left(\frac{\tau-\rho}{\tau+\rho}\right)^{1/4} \end{pmatrix}. \tag{5.35}$$

The solutions using the zweibein set e_A Eq. (5.32) is

$$\psi_e(\tau, \rho) = L\, \psi_h(\tau, \rho), \tag{5.36}$$

where $\psi_h(\tau, \rho)$ is given by Eq. (5.31). We have then,

$$\psi_e(\tau, \rho) = e^{-ip_x \tanh^{-1}\left(\frac{\rho}{\tau}\right)} \begin{pmatrix} \left(\frac{\tau+\rho}{\tau-\rho}\right)^{1/4} (A\, J_\nu(z) + B\, Y_\nu(z)) \\ i\left(\frac{\tau-\rho}{\tau+\rho}\right)^{1/4} (A\, J_{\nu-1}(z) + B\, Y_{\nu-1}(z)) \end{pmatrix}, \tag{5.37}$$

where

$$\nu = \frac{1}{2} - \frac{ip_x}{a_0}, \tag{5.38}$$

$$z = m\sqrt{\tau^2 - \rho^2}. \tag{5.39}$$

The wavefunction $\psi_e(\tau, \rho)$, satisfies Eq. (5.4) which now reduces to the usual Minkowski spacetime Dirac equation, namely,

$$\left(i\gamma^A \partial_A - m I_2\right) \psi_e = 0. \tag{5.40}$$

In order to show that $\psi_e(\tau, \rho)$ satisfies Eq. (5.40), one has to use the Bessel function identity,

$$C_{\nu-1}(z) + C_{\nu+1}(z) = \frac{2\nu}{z} C_\nu(z), \tag{5.41}$$

where $C_\nu(z)$ denotes either of the Bessel functions $J_\nu(z)$, $Y_\nu(z)$.

(c) In this subsection we shall consider the normalization integral [23], p. 69. This integral is referred to as the "probability integral" in [24]. Thus in Fermi coordinates with the canonical zweibein we have,

$$(\psi_e \mid \psi_e) = \int_\Sigma \bar{\psi}_e \, \gamma^0 \psi_e \, d\rho. \tag{5.42}$$

In our case the spacelike hypersurface Σ is the usual (τ_0, ρ), $\tau_0 > 0$, hyperplane. Using Eq. (5.7) we have

$$(\psi_e \mid \psi_e) = \int_{-\tau_0}^{\tau_0} \psi_e^\dagger \, \psi_e \, d\rho. \tag{5.43}$$

For the remainder of this section we will write $\bar{a}$ to denote the complex conjugate of a. We shall make use of the fact that

$$\overline{C_\nu(z)} = C_{\bar{\nu}}(\bar{z}), \tag{5.44}$$

where again $C_\nu(z)$ denotes either of the Bessel functions $J_\nu(z)$, $Y_\nu(z)$. Since in our case z is real we have that

$$\overline{C_\nu(z)} = C_{\bar{\nu}}(z). \tag{5.45}$$

In order to check the behavior of the integrand in Eq. (5.43) at the endpoints of integration, we use the limiting forms of the Bessel functions when ν is fixed and $z \sim 0$. Using Abramowitz and Stegun's Eqs. (9.1.2), p. 358 and (9.1.7), p. 360, [25], one easily deduces that for $\nu \neq$ negative integer,

$$Y_\nu(z) = \frac{J_\nu(z) \cos(\nu\pi) - J_{-\nu}(z)}{\sin(\nu\pi)}, \tag{5.46}$$

$$J_\nu(z) \sim \left(\frac{z}{2}\right)^\nu \frac{1}{\Gamma(\nu+1)}, \tag{5.47}$$

$$Y_\nu(z) \sim -\left(\frac{z}{2}\right)^{-\nu} \frac{\Gamma(\nu)}{\pi}, \qquad \mathrm{Re}(\nu) > 0, \tag{5.48}$$

$$Y_\nu(z) \sim -\left(\frac{z}{2}\right)^\nu \frac{\cot(\nu\pi)}{\Gamma(\nu+1)}, \qquad \mathrm{Re}(\nu) < 0. \tag{5.49}$$

Using Eqs. (5.47)–(5.49) in the integrand of Eq. (5.43), we see that some terms blow up at the endpoints like $\sim 1/(\tau_0 - \rho)$ as the integration variable $\rho \to \tau_0$, regardless of the value of p_x. We can eliminate these terms by setting $B = 0$. Thus the solution ψ_e of Eq. (5.37) reduces to

$$\psi_e(\tau, \rho) = A\, e^{-ip_x \tanh^{-1}\left(\frac{\rho}{\tau}\right)} \begin{pmatrix} \left(\frac{\tau+\rho}{\tau-\rho}\right)^{1/4} J_\nu(z) \\ i\left(\frac{\tau-\rho}{\tau+\rho}\right)^{1/4} J_{\nu-1}(z) \end{pmatrix}, \qquad (5.50)$$

where A is a normalization constant. Going back to the original coordinates Eq. (5.14) we may write

$$\psi(t, x) = A\, e^{-ip_x x} \begin{pmatrix} J_\nu(mt) \\ i\, J_{\nu-1}(mt) \end{pmatrix}. \qquad (5.51)$$

There is also the solution with the opposite sign of p_x, [14], (their k is the negative of our p_x)

$$\psi(t, x) = A\, e^{ip_x x} \begin{pmatrix} J_{-\nu+1}(mt) \\ i\, J_{-\nu}(mt) \end{pmatrix}. \qquad (5.52)$$

References

1. C. Teitelboim, The Hamiltonian structure of two-dimensional space-time and its relation with the conformal anomaly, in *Quantum Theory of Gravity*, ed. by S.M. Christensen (Adam Hilger, Bristol 1984), pp. 327–344
2. R. Jackiw, Liouville field theory: a two-dimensional model for gravity?, in *Quantum Theory of Gravity*, ed. by S.M. Christensen (Adam Hilger, Bristol, 1984), pp. 403–420,
3. J.D. Brown, M. Henneaux, C. Teitelboim, Black holes in two spacetime dimensions. Phys. Rev. D **33**, 319–323 (1986)
4. R.B. Mann, A. Shiekh, L. Tarasov, Classical and quantum properties of two-dimensional black holes. Nuclear Phys. B **341**, 134–154 (1990)
5. R.B. Mann, The simplest black holes. Found. Phys. Letters **4**, 425–449 (1991)
6. E. Witten, String theory and black holes. Phys. Rev. D **44**, 314–324 (1991)
7. R.B. Mann, Lower dimensional black holes. Gen. Relativ. Gravit. **24**, 433–449 (1992)
8. J.D. Brown, *Lower Dimensional Gravity* (World Scientific, Singapore, 1988)
9. N.D. Birrell, P.C.W. Davies, *Quantum Fields in Curved Space* (Cambridge U. Press, New York, 1984)
10. J. Gegenberg, P.F. Kelly, R.B. Mann, D. Vincent, Theories of gravitation in two dimensions. Phys. Rev. D **37**, 3463–3471 (1988)
11. E. Brézin, S.R. Wadia (eds.), *The Large N Expansion in Quantum Field Theory and Statistical Physics, from Spin Systems to 2-Dimensional Gravity* (World Scientific, Singapore, 1993)
12. R.B. Mann, S.M. Morsink, A.E. Sikkema, T.G. Steele, Semiclassical gravity in $(1 + 1)$ dimensions. Phys. Rev. D **43**, 3948–3957 (1991)
13. S.M. Morsink, R.B. Mann, Black hole radiation of Dirac particles in $1 + 1$ dimensions. Class. Quantum Grav. **8**, 2257–2268 (1991)
14. A. Sinha, R. Roychoudhury, Dirac equation in $(1 + 1)$-dimensional curved space-time. Int. J. Theor. Phys. **33**, 1511–1522 (1994)
15. R. Bécar, P.A. González, Y. Vásquez, Dirac quasinormal modes of two-dimensional charged dilatonic black holes. Eur. Phys. J. C **74**(2940), 1–5 (2014)

16. Ö. Yeşiltaş, Exact solutions of the Dirac Hamiltonian on the sphere under hyperbolic magnetic fields. Adv. High Energy Phys. **2014**, 186425, (2014), https://www.hindawi.com/journals/ahep/2014/186425/abs/

17. L. Gosse, Locally inertial approximations of balance laws arising in $(1 + 1)$-dimensional general relativity. SIAM J. Appl. Math. (SIAP) **75**, 1301–1328 (2015), https://hal.archives-ouvertes.fr/hal-01119168

18. S.K. Moayedi, F. Darabi, Exact solutions of Dirac equation on a 2D gravitational background. Phys. Lett. A. **322**, 173–178 (2004)

19. P.B. Pal, Dirac, Majorana, and Weyl fermions. Am. J. Phys. **79**, 485–498 (2011)

20. D. Klein, E. Randles, Fermi coordinates, simultaneity, and expanding space in Robertson-Walker cosmologies. Ann. Henri Poincaré **12**, 303–328 (2011)

21. D. Klein, J. Reschke, Pre-big bang geometric extensions of inflationary cosmologies. Ann. Henri Poincaré **19**, 565–606 (2018)

22. P. Collas, D. Klein, Dirac particles in a gravitational shock wave. Class. Quantum Grav. **35**, 125006 (2018), https://doi.org/10.1088/1361-6382/aac144

23. C. Itzykson, J.-B. Zuber, *Quantum Field Theory* (McGraw-Hill Inc., New York, 1980)

24. F. Finster, M. Reintjes, The Dirac equation and the normalization of its solutions in a closed Friedmann-Robertson-Walker universe. Class. Quantum Grav. **26**, 105021 (2009), arXiv:0901.0602v4

25. M. Abramowirz, I.A. Stegun, *Handbook of Mathematical Functions*, (National Bureau of Standards, Applied Math. Series # 55, Washington D. C., 1972)

Chapter 6
The Dirac Equation in $(2+1)$ GR

6.1 Introduction to $(2+1)$

Gravity and quantum gravity in $(2+1)$ spacetime dimensions has been a very active and useful research area as the following four texts attest: *Lower Dimensional gravity* by Brown [1], *Quantum Gravity in* $2+1$ *Dimensions by Carlip* [2], *Exact Solutions in Three-Dimensional Gravity* by García-Díaz [3] and Chap. 2 in *Advanced Lectures in General Relativity* by Compére and Fiorucci [4]. Solutions of the gravitational field equations in $(2+1)$ spacetimes were investigated by several authors, [5, 6]. Furthermore apart from investigations of solutions of the Dirac equation in gravitational spacetimes [7–9], the formalism of the massless Dirac equation in $(2+1)$ dimensions has been successfully applied in solid state physics, in particular for describing "curved graphene," and fullerenes, [10–14], and Dirac semi-metals, [15]. We recommend the review by Iorio [16].

The Dirac equation in $(2+1)$ Minkowski spacetime has been discussed in several papers [17, 18]. An important reference is the paper by Gavrilov et al. [18], especially if one wishes to introduce an electromagnetic vector potential A_C. We shall follow their conventions.

In $(2+1)$ dimensions the Dirac spinors are two-component spinors and one can write a set of 2×2 γ-matrices satisfying all the required conditions using the Pauli matrices

$$\sigma^1 = \begin{pmatrix} 0 & 1 \\ 1 & 0 \end{pmatrix}, \quad \sigma^2 = \begin{pmatrix} 0 & -i \\ i & 0 \end{pmatrix}, \quad \sigma^3 = \begin{pmatrix} 1 & 0 \\ 0 & -1 \end{pmatrix}. \tag{6.1}$$

We adopt the metric signature $(+, -, -)$ and choose the following set of constant 2×2 Dirac matrices,[1]

$$\gamma^0 = \sigma^3, \quad \gamma^1 = i\sigma^2, \quad \gamma^2 = -i\sigma^1. \tag{6.2}$$

---

[1] This choice corresponds to $s = +1$ in Ref. [18], while Ref. [17] uses the negatives of our Eq. (6.1).

© The Author(s), under exclusive license to Springer Nature Switzerland AG 2019
P. Collas and D. Klein, *The Dirac Equation in Curved Spacetime*,
SpringerBriefs in Physics, https://doi.org/10.1007/978-3-030-14825-6_6

Thus

$$\left(\gamma^0\right)^2 = I_2, \quad \left(\gamma^1\right)^2 = \left(\gamma^2\right)^2 = -I_2. \tag{6.3}$$

We note that, as usual, they satisfy the Hermiticity conditions

$$(\gamma^A)^\dagger = \gamma^0 \gamma^A \gamma^0, \quad A = 0, 1, 2. \tag{6.4}$$

With the above choices the chirality operator is simply $\gamma^5 := \gamma^0$.

For a free spin $1/2$ particle of mass m the Dirac equation in Minkowski spacetime is

$$i\gamma^A \partial_A \psi - m\psi = 0, \tag{6.5}$$

where ψ is a 2-spinor.

6.2   The Dirac Equation in the Schrödinger Universe

The Schrödinger universe [19] is spatially flat but it is not an empty spacetime solution of the field equations. The Dirac equation in this spacetime was solved by Barut and Duru in a paper dealing with spatially flat $(3 + 1)$ Robertson-Walker cosmologies [20]. In $(2 + 1)$ the metric is,

$$ds^2 = dt^2 - (a_0 t)^2 \left(dx^2 + dy^2\right). \tag{6.6}$$

The vielbein (dreibein) 1-form fields and corresponding vectors are given below.

$$e^0 = dt, \qquad\qquad\qquad e_0 = \partial_t, \tag{6.7}$$

$$e^1 = a_0 t\, dx, \qquad\qquad e_1 = \frac{1}{a_0 t} \partial_x, \tag{6.8}$$

$$e^2 = a_0 t\, dy, \qquad\qquad e_2 = \frac{1}{a_0 t} \partial_y. \tag{6.9}$$

The spacetime dependent matrices $\bar{\gamma}^\mu$ are given, as usual, by

$$\bar{\gamma}^\mu = e_A{}^\mu \gamma^A, \tag{6.10}$$

where

$$g_{\mu\nu} = e^A{}_\mu e^B{}_\nu \eta_{AB}. \tag{6.11}$$

For the rest of the chapter, we shall adopt our conventions for the spinor connection coefficients, Eqs. 3.24, 3.26, 3.27 adapted to $(2 + 1)$. We outline the steps involved and the results obtained below.

Using Eq. (6.10) we find that

$$\bar{\gamma}^t = \gamma^0, \tag{6.12}$$

$$\bar{\gamma}^x = \frac{1}{a_0 t} \gamma^1, \tag{6.13}$$

$$\bar{\gamma}^y = \frac{1}{a_0 t} \gamma^2. \tag{6.14}$$

The nonvanishing affine connection coefficients are

$$\Gamma^t{}_{xx} = \Gamma^t{}_{yy} = a_0^2 t, \tag{6.15}$$

$$\Gamma^x{}_{tx} = \Gamma^y{}_{ty} = \frac{1}{t}. \tag{6.16}$$

The nonvanishing covariant derivatives of the tetrad vector fields, $e_A{}^\alpha$, are

$$\nabla_x e_1{}^t = \nabla_y e_2{}^t = a_0, \tag{6.17}$$

$$\nabla_x e_0{}^x = \nabla_y e_0{}^y = \frac{1}{t}. \tag{6.18}$$

The nonvanishing $\omega_{AB\mu}$, are

$$\omega_{01x} = \omega_{02y} = a_0. \tag{6.19}$$

Finally, the spinor affine connection coefficients Γ_μ are given by

$$\Gamma_t = 0, \tag{6.20}$$

$$\Gamma_x = \frac{1}{2} \omega_{01x} \gamma^0 \gamma^1, \tag{6.21}$$

$$\Gamma_y = \frac{1}{2} \omega_{02y} \gamma^0 \gamma^2. \tag{6.22}$$

It is convenient to use the Dirac equation in the form given in Eq. 3.31, namely,

$$i\bar{\gamma}^\mu (\partial_\mu + \Gamma_\mu) \psi - m\psi = 0. \tag{6.23}$$

From the form of the metric, Eq. (6.6), it is clear that we may write the spinor ψ in form

$$\psi(t, x, y) = \begin{pmatrix} \psi_1(t, x, y) \\ \psi_2(t, x, y) \end{pmatrix} = e^{-i(p_x x + p_y y)} \begin{pmatrix} F(t) \\ G(t) \end{pmatrix}. \tag{6.24}$$

We now use Eqs. (6.12)–(6.14), along with Eqs. (6.20)–(6.22), in the Dirac equation (6.23) and substitute ψ from Eq. (6.24). We then obtain the two differential equations below for the time-dependent factors of the spinor components, $F(t)$ and

$G(t)$, where we have excluded the point $t = 0$.

$$a_0 t F'(t) + a_0 (1 + imt) F(t) - (ip_x + p_y) G(t) = 0, \tag{6.25}$$
$$a_0 t G'(t) + a_0 (1 - imt) G(t) - (ip_x - p_y) F(t) = 0. \tag{6.26}$$

We solve Eq. (6.25) for $G(t)$ and substitute the result in Eq. (6.26) and obtain the differential equation for $F(t)$ below,

$$F''(t) + \frac{3}{t} F'(t) + \frac{(1 + k^2 + mt(i + mt))}{t^2} F(t) = 0, \tag{6.27}$$

where the constant k^2 is defined by

$$k^2 = \frac{p_x^2 + p_y^2}{a_0^2}. \tag{6.28}$$

We shall choose

$$k = \frac{\sqrt{p_x^2 + p_y^2}}{a_0}, \quad a_0 > 0. \tag{6.29}$$

Following the standard procedure for eliminating the F' from Eq. (6.27), we define the function

$$q(t) = \frac{1 + 4 (k^2 + imt + m^2 t^2)}{4t^2}, \tag{6.30}$$

and obtain the differential equation

$$u''(t) + q(t)u(t) = 0, \tag{6.31}$$

where

$$F(t) = t^{-\frac{3}{2}} u(t). \tag{6.32}$$

We find the solution

$$F(t) = t^{-\frac{3}{2}} \left[A \, W_{\frac{1}{2}, ik}(z) + B \, M_{\frac{1}{2}, ik}(z) \right], \tag{6.33}$$

where $M_{\kappa,\mu}(z)$ and $W_{\kappa,\mu}(z)$ are, respectively, Whittaker's functions of the first and second kind and A and B are arbitrary constants. Using the recurrence relations found in pp. 303–304 of [21], we may write $G(t)$ as

$$G(t) = Ct^{-\frac{3}{2}} \left[-ikA \, W_{-\frac{1}{2}, ik}(z) + B \, M_{\frac{1}{2}, ik}(z) \right], \tag{6.34}$$

where the dimensionless constant C is given by

$$C = \frac{p_x + ip_y}{a_0 k} = \frac{p_x + ip_y}{\sqrt{p_x^2 + p_y^2}} = -\frac{p^x + ip^y}{\sqrt{(p^x)^2 + (p^y)^2}} . \tag{6.35}$$

The complete spinor ψ is given by substituting the above F and G into Eq. (6.24) and is an eigenstate of the canonical momenta p_x and p_y. We can gain more insight in the solution given by (6.33) and (6.34), by going to the rest frame of the Dirac particle, namely, $k = 0$. From Ref. [21], p. 305, we have the relations below:

$$W_{\frac{1}{2},0}(2imt) = M_{\frac{1}{2},0}(2imt) = e^{-imt} \sqrt{2imt} . \tag{6.36}$$

$$M_{-\frac{1}{2},0}(2imt) = e^{imt} \sqrt{2imt} . \tag{6.37}$$

Therefore the component $F(t)$, is well behaved and involves positive energies only.[2] However there are two problems in connection with $G(t)$. One can see by looking at the large t asymptotic behavior of W and M, that M contains both positive and negative energy contributions, [22] (13.19.2), as opposed to W, which for large argument, $2imt = z \to \infty$, $|\text{ph } z| = \frac{\pi}{2} \leq \frac{3\pi}{2} - \delta$, behaves as [23]

$$W_{\frac{1}{2},k}(2imt) \sim e^{-imt} \sqrt{2imt} , \tag{6.38}$$

$$W_{-\frac{1}{2},k}(2imt) \sim e^{-imt} \frac{1}{\sqrt{2imt}} . \tag{6.39}$$

The second problem is that the value of the constant C, Eq. (6.35), in the rest frame, depends on the way we reach that limit, i.e., whether we let p_x or p_y go to zero first. Both problems can be eliminated by letting the constant $B = 0$ in $F(t)$ and $G(t)$. Thus now we have the positive energy spinor

$$\psi(t, x, y) = e^{-i(p_x x + p_y y)} \begin{pmatrix} A \\ t^{\frac{3}{2}} \end{pmatrix} \begin{pmatrix} W_{\frac{1}{2}, ik}(z) \\ -i \frac{p_+}{a_0} W_{-\frac{1}{2}, ik}(z) \end{pmatrix} , \tag{6.40}$$

where $p_+ = p_x + ip_y$. Thus in the rest frame of the particle, where $p_+ = 0$, we obtain

$$\psi(t, x, y) = \psi(t) = \begin{pmatrix} \frac{A\sqrt{2im}}{t} \end{pmatrix} \begin{pmatrix} e^{-imt} \\ 0 \end{pmatrix} . \tag{6.41}$$

Remark 14   The solutions obtained by setting $p_x = p_y = 0$ and solving the resulting Eqs. (6.25) and (6.26), confirm our choices above and the result in Eq. (6.41).

To find the negative energy spinor one has to repeat the whole process but this time use the exponential factor $e^{+i(p_x x + p_y y)}$.

---

[2] One could also consult Sect. 13, (in particular 13.14.10), of the excellent website DLMF [22].

References

1. J.D. Brown, *Lower Dimensional Gravity* (World Scientific, Singapore, 1988)
2. S. Carlip, *Quantum Gravity in 2 + 1 Dimensions* (Cambridge U. Press, Cambridge, 1998)
3. A.A. García-Díaz, *Exact Solutions in Three-Dimensional Gravity* (Cambridge U. Press, Cambridge, 2017)
4. G. Compère, A. Fiorucci, *Advanced Lectures on General Relativity* (2018), arXiv:1801.07064v3
5. P. Collas, General relativity in two- and three-dimensional space-times. Am. J. Phys. **45**, 833–837 (1977)
6. A. Corichi, A. Gomberoff, On a spacetime duality in 2 + 1 gravity. Class. Quantum Grav. **16**, 3579–3598 (1999)
7. Y. Sucu, N. Ünal, Exact solution of the Dirac equation in 2 + 1 dimensional gravity. J. Math Phys. **48**, 052503 (2007)
8. Ö. Yeşiltaş, Non-Hermitian Dirac Hamiltonian in three-dimensional gravity and pseudosupersymmetry. Adv. High Energy Phys. **2015**, 484151, (2015), https://www.hindawi.com/journals/ahep/2015/484151/abs/
9. Ö. Yeşiltaş, Dirac equation on the torus and rationally extended trigonometric potentials within supersymmetric QM, pp. 1–12. 19 July 2017. arXiv:1707.06136v1
10. D.P. DiVincenzo, E.J. Mele, Self-consistent effective-mass theory for intralayer screening in graphite intercalation compounds. Phys. Rev. B **29**, 1685–1694 (1984)
11. V.A. Osipov, D.V. Kolesnikov, Electronic properties of curved carbon nanostructures. Rom. Journ. Phys. **50**, 457–466 (2005)
12. M.A.H. Vozmediano, M.I. Katsnelson, F. Guinea, Gauge fields in graphene. Phys. Rep. **496**, 109–148 (2010)
13. M.B. Belonenko, N.G. Lebedev, N.N. Yanyushkina, A.V. Zhukov, M. Paliy, Electronic spectrum and tunneling current in curved graphene nanoribbons. Solid State Comm. **151**, 1147–1150 (2011)
14. P. Kosiński, P. Maślanka, J. Slawińska, I. Zasada, QED_{2+1} in graphene: symmetries of dirac equation in 2 + 1 dimensions. Prog. Theoret. Phys. **128**, 727–739 (2012)
15. M. Rogatko, K.I. Wysokinski, Hydrodynamics of topological Dirac semi-metals with chiral and Z_2 anomalies. arXiv:1804.02202v2
16. A. Iorio, Curved spacetimes and curved graphene: a status report of the weyl symmetry approach. Int. J. Mod. Phys. D **24**(1530013), 1–63 (2015)
17. R. Jackiw, V.P. Nair, Relativistic wave equation for anyons. Phys. Rev. D **43**, 1933–1942 (1991)
18. S.P. Gavrilov, D.M. Gitman, J.L. Tomazelli, Comments on spin operators and spin-polarization states of 2 + 1 fermions. Eur. Phys. J. **39**, 245–248 (2005)
19. E. Schrödinger, *Expanding Universes* (Cambridge U. Press, Cambridge, 1957)
20. A.O. Barut, I.H. Duru, Exact solutions of the Dirac equation in spatially flat Robertson-Walker space-times. Phys. Rev. D **36**, 3705–3711 (1987)
21. W. Magnus, F. Oberhettinger, R.P. Soni, *Formulas and Theorems for the Special Functions of Mathematical Physics*, 3rd edn. (Springer-Verlag, Berlin, 1966)
22. *Digital Library of Mathematical Functions*. https://dlmf.nist.gov/
23. F.W.J. Olver, *Asymptotics and Special Functions* (A. K. Peters, Wellesley, 1997)

Chapter 7
Scalar Product

7.1 Conservation of j Special Relativity

The probability current density for a Dirac field is given by

$$j^A = \bar{\psi}\gamma^A\psi, \tag{7.1}$$

where the adjoint spinor is $\bar{\psi} = \psi^\dagger\gamma^0$. The probability current density j^A transforms like a 4-vector under a Lorentz transformation, Λ, so that,

$$j'^A = \Lambda^A{}_B\, j^B, \tag{7.2}$$

moreover j^A is conserved, that is,

$$\partial_A j^A = 0. \tag{7.3}$$

First we will prove Eq. (7.2). We know that under a tetrad rotation (local Lorentz transformation) Λ,

$$\psi' = L\psi \quad \Rightarrow \quad \psi'^\dagger = \psi^\dagger L^\dagger, \tag{7.4}$$

where L and Λ are related as in (3.6). Thus, using Eqs. (A.37) and (A.38), we have that

$$j'^A = \psi'^\dagger\gamma^0\gamma^A\psi', \tag{7.5}$$
$$= \psi^\dagger L^\dagger\gamma^0\gamma^A L\psi \tag{7.6}$$
$$= \psi^\dagger\gamma^0\left(L^{-1}\gamma^A L\right)\psi, \tag{7.7}$$
$$= \psi^\dagger\gamma^0\Lambda^A{}_B\gamma^B\psi, \tag{7.8}$$
$$= \Lambda^A{}_B\psi^\dagger\gamma^0\gamma^B\psi, \tag{7.9}$$
$$= \Lambda^A{}_B\, j^B. \tag{7.10}$$

© The Author(s), under exclusive license to Springer Nature Switzerland AG 2019
P. Collas and D. Klein, *The Dirac Equation in Curved Spacetime*,
SpringerBriefs in Physics, https://doi.org/10.1007/978-3-030-14825-6_7

Next in order to show that the current is conserved it will be useful to write the expression for the adjoint of the Dirac equation. We begin with the usual special relativity Dirac equation,

$$i\gamma^A \partial_A \psi - m\psi = i\partial\!\!\!/\,\psi - m\psi = 0. \tag{7.11}$$

Then the adjoint is obtained as follows:

$$\left(i\gamma^A \partial_A \psi - m\psi\right)^\dagger = 0, \tag{7.12}$$

$$-i\partial_A \psi^\dagger \left(\gamma^A\right)^\dagger - m\psi^\dagger = 0, \tag{7.13}$$

$$i\partial_A \psi^\dagger \gamma^0 \gamma^A \gamma^0 + m\psi^\dagger = 0, \tag{7.14}$$

$$i\left(\partial_A \bar{\psi}\right)\gamma^A + m\bar{\psi} = 0, \tag{7.15}$$

where we used the relations (B.4) and (B.12). Thus we shall use the notation

$$\bar{\psi}\overleftarrow{\partial\!\!\!/} = \partial_A \bar{\psi}\gamma^A, \tag{7.16}$$

$$\overrightarrow{\partial\!\!\!/}\psi = \partial_A \psi\gamma^A. \tag{7.17}$$

So we have the shorthand Feynman slash notation for the Dirac equation and its adjoint:

$$\left(i\overrightarrow{\partial\!\!\!/} - mI\right)\psi = 0, \tag{7.18}$$

$$\bar{\psi}\left(i\overleftarrow{\partial\!\!\!/} + mI\right) = 0. \tag{7.19}$$

Multiplying Eq. (7.18) on the left with $\bar{\psi}$, and Eq. (7.19) on the right with ψ and adding, we obtain

$$\bar{\psi}\left(\overleftarrow{\partial\!\!\!/} + \overrightarrow{\partial\!\!\!/}\right)\psi \equiv \partial_A\left(\bar{\psi}\gamma^A\psi\right) = 0, \tag{7.20}$$

which completes the proof of the conservation Eq. (7.3).

It follows from Eq. (7.3) that

$$\frac{d}{dt}\int_V j^0 d^3x = -\int_V \partial_K j^K d^3x = -\int_{\partial V} j^K dS_K = 0, \quad K = (1,2,3). \tag{7.21}$$

In Eq. (7.21) we have used Gauss' theorem where ∂V is the boundary of the volume V, so that we may write $j^K dS_K = j^1 dx^2 \wedge dx^3 + j^2 dx^3 \wedge dx^1 + j^3 dx^1 \wedge dx^2$. The last step of Eq. (7.21) is valid for infinite volumes, with surface at infinity, provided ψ vanishes sufficiently fast there. Equation (7.21) *is an expression of conservation of (total) probability in time.*

7.2 The Current Density in General Relativity

The probability current density in general relativity (curved spacetime) is given by

$$j^\alpha = \bar\psi \, \bar\gamma^\alpha(x)\psi, \qquad (7.22)$$

where ψ is a solution of Eq. (3.31), the $\bar\gamma^\alpha(x)$ are given by Eq. (3.3) and $\bar\psi = \psi^\dagger \gamma^0$. The curved spacetime proof of Eq. (7.3) is given in [1, 2], p. 145, and follows similar steps as the above derivation except that partial derivatives become covariant derivatives and, of course, one has to use Eq. (3.31) instead of the Minkowski spacetime Dirac equation. A discussion of the generalization of Eq. (7.21) is given in Appendix E in Ref. [3]. We also refer the reader to our Proposition 1.

7.3 The Scalar Product in Special Relativity

We now define the usual scalar product (discussed in more detail in Appendix D.3)

$$(\phi \,|\, \psi) = \int_\Sigma \bar\phi \gamma^0 \psi \, d^3 x \, . \qquad (7.23)$$

In the case where our spacelike hypersurface Σ is not the usual (t_0, x, y, z) hyperplane, Eq. (7.23) generalizes to

$$(\phi \,|\, \psi) = \int_\Sigma \bar\phi \gamma^A n_A \, \psi \, d\Sigma \, , \qquad (7.24)$$

where n is the future-directed normal to Σ, and $d\Sigma$ is the invariant "volume element" on Σ. The probability integral $(\psi|\psi)$ is then given by

$$(\psi \,|\, \psi) = \int_\Sigma j^A n_A \, d\Sigma \, . \qquad (7.25)$$

In special relativity one usually chooses Σ to be the $t = 0$ hyperplane.

7.4 The Scalar Product in General Relativity

We follow Ref. [4] and define the scalar product

$$(\phi \mid \psi) = \int_{\Sigma} \bar{\phi} \, \bar{\gamma}^{\alpha}(x) n_{\alpha} \psi \, d\Sigma \,, \tag{7.26}$$

where $\bar{\phi} = \phi^{\dagger} \gamma^{0}$, and the $\bar{\gamma}^{\alpha}(x)$ are given by Eq. (3.3). The vector n is the future-directed normal to the spacelike Cauchy hypersurface Σ, and $d\Sigma$ is the invariant "volume element" on Σ. Using Eq. (7.22) we have that the probability integral $(\psi|\psi)$ is given by

$$(\psi \mid \psi) = \int_{\Sigma} j^{\alpha} n_{\alpha} \, d\Sigma \,. \tag{7.27}$$

We briefly comment on Parker's definitions [1, 5]. Parker defines the current of Eq. (7.22) with a minus sign in front. This is necessary because he has chosen a representation where $\left(\gamma^{0}\right)^{2} = -I$ and metric the signature $(-, +, +, +)$ (see also his argument regarding the positive definiteness of $(\psi \mid \psi)$ around his Eq. (3.5)). In addition he defines the scalar product

$$(\phi \mid \psi) = -\int_{\Sigma} \bar{\phi} \, \bar{\gamma}^{0}(x) \psi \, \sqrt{-g} \, d^{3}x \,, \tag{7.28}$$

where the integration is over a constant x^{0} Cauchy hypersurface ($d^{3}x = d\Sigma_{t}$). Parker is essentially using the *lapse and shift* formulation [6], where the metric is written in the $(3 + 1)$ decomposition (using Parker's [1] and Poisson's [6] $(+2)$ signature, cf. Sect. 3.2)

$$ds^{2} = -N^{2}dt^{2} + h_{ab} \left(dx^{a} + N^{a}dt\right) \left(dx^{b} + N^{b}dt\right) \,. \tag{7.29}$$

In Eq. (7.29) N and N^{a} are the lapse and shift functions respectively, h_{ab} is the induced metric on Σ_{t} and one can show that

$$\sqrt{-g} = N\sqrt{h}. \tag{7.30}$$

Therefore definitions (7.26) and (7.28) agree.

We would now like to prove Proposition 1.

Proposition 1
$$D_{\mu} \left(\bar{\phi} \, \bar{\gamma}^{\mu} \psi\right) = 0. \tag{7.31}$$

The proof of this Proposition is simple provided one has certain preliminary results available. So we first derive the required results.

We give the derivation implied in Ref. [7], Eq. (21), in order to find the expression for $D_\mu\bar{\psi}$, where ψ is a solution of the Dirac equation. We use the fact that $\bar{\psi}\psi$ is a 0-form field, therefore

$$D_\mu\left(\bar{\psi}\psi\right) = \left(D_\mu\bar{\psi}\right)\psi + \bar{\psi}D_\mu\psi\,, \tag{7.32}$$

$$= \left(D_\mu\bar{\psi}\right)\psi + \bar{\psi}\left(I\partial_\mu + \Gamma_\mu\right)\psi\,, \tag{7.33}$$

$$\equiv \left(I\partial_\mu\bar{\psi} + \mathbb{G}_\mu\bar{\psi}\right)\psi + \bar{\psi}\partial_\mu\psi + \bar{\psi}\Gamma_\mu\psi\,, \tag{7.34}$$

$$= \left(\partial_\mu\bar{\psi}\right)\psi + \bar{\psi}\partial_\mu\psi\,. \tag{7.35}$$

Equations (7.34) and (7.35) imply that

$$\left(\mathbb{G}_\mu\bar{\psi}\right)\psi + \bar{\psi}\Gamma_\mu\psi = 0\,, \tag{7.36}$$

$$\left(\mathbb{G}_\mu\bar{\psi} + \bar{\psi}\Gamma_\mu\right)\psi = 0\,, \tag{7.37}$$

$$\mathbb{G}_\mu\bar{\psi} + \bar{\psi}\Gamma_\mu = 0\,. \tag{7.38}$$

Thus

$$\mathbb{G}_\mu\bar{\psi} = -\bar{\psi}\Gamma_\mu\,, \tag{7.39}$$

so, finally, using Eqs. (7.33), (7.34), and (7.39) we may write

$$D_\mu\bar{\psi} = I\partial_\mu\bar{\psi} - \bar{\psi}\Gamma_\mu\,. \tag{7.40}$$

Now we would like to find the adjoint of the Dirac equation in curved spacetime, i.e.,

$$\left(\bar{\gamma}^\mu D_\mu\psi + im\psi\right)^\dagger = 0\,. \tag{7.41}$$

The derivation requires three steps. The first step to note that

$$(\bar{\gamma}^\mu)^\dagger = e_A{}^\mu\left(\gamma^A\right)^\dagger = e_A{}^\mu\gamma^0\gamma^A\gamma^0 = \gamma^0\bar{\gamma}^\mu\gamma^0\,. \tag{7.42}$$

The second step is to recall the expression for Γ_μ and obtain the result below

$$\Gamma_\mu^\dagger = \frac{1}{4}\omega_{AB\mu}\left(\gamma^A\gamma^B\right)^\dagger\,, \tag{7.43}$$

$$= \frac{1}{4}\omega_{AB\mu}\left(\gamma^B\right)^\dagger\left(\gamma^A\right)^\dagger\,, \tag{7.44}$$

$$= \frac{1}{4}\omega_{AB\mu}\left(\gamma^0\gamma^B\gamma^0\gamma^0\gamma^A\gamma^0\right)\,, \tag{7.45}$$

$$= \frac{1}{4}\omega_{AB\mu}\left(\gamma^0\gamma^B\gamma^A\gamma^0\right)\,, \tag{7.46}$$

$$= -\frac{1}{4}\omega_{AB\mu}\left(\gamma^0\gamma^A\gamma^B\gamma^0\right)\,, \tag{7.47}$$

$$= -\gamma^0\Gamma_\mu\gamma^0\,, \tag{7.48}$$

In the third and final step we make use of Eqs. (7.42) and (7.48) and re-write Eq. (7.41) as follows,

$$\left(D_\mu \psi\right)^\dagger (\bar\gamma^\mu)^\dagger - im\psi^\dagger = 0, \tag{7.49}$$

$$\left[(I\partial_\mu + \Gamma_\mu)\,\psi\right]^\dagger (\bar\gamma^\mu)^\dagger - im\psi^\dagger = 0, \tag{7.50}$$

$$\left(\partial_\mu \psi^\dagger + \psi^\dagger \Gamma_\mu^\dagger\right) (\bar\gamma^\mu)^\dagger - im\psi^\dagger = 0, \tag{7.51}$$

$$\left(\partial_\mu \psi^\dagger\right) \gamma^0 \bar\gamma^\mu \gamma^0 - \psi^\dagger \gamma^0 \Gamma_\mu \gamma^0 \gamma^0 \bar\gamma^\mu \gamma^0 - im\psi^\dagger = 0, \tag{7.52}$$

$$\left(\partial_\mu \bar\psi\right) \bar\gamma^\mu \gamma^0 - \bar\psi \Gamma_\mu \bar\gamma^\mu \gamma^0 - im\psi^\dagger = 0, \tag{7.53}$$

$$\left(\partial_\mu \bar\psi\right) \bar\gamma^\mu - \bar\psi \Gamma_\mu \bar\gamma^\mu - im\bar\psi = 0, \tag{7.54}$$

$$\left(I\partial_\mu \bar\psi - \bar\psi \Gamma_\mu\right) \bar\gamma^\mu - im\bar\psi = 0, \tag{7.55}$$

$$\left(D_\mu \bar\psi\right) \bar\gamma^\mu - im\bar\psi = 0. \tag{7.56}$$

Finally, we return to the proof of Proposition 1. We assume that ϕ and ψ satisfy the Dirac Eq. (3.2) and its adjoint Eq. (7.56), namely,

$$i\bar\gamma^\mu D_\mu \psi - m\psi = 0, \quad i\left(D_\mu \bar\psi\right) \bar\gamma^\mu + m\bar\psi = 0, \tag{7.57}$$

from which it follows that,

$$\bar\gamma^\mu D_\mu \psi = -im\psi, \quad \left(D_\mu \bar\psi\right) \bar\gamma^\mu = im\bar\psi. \tag{7.58}$$

The proof of Proposition 1 now follows by applying the Leibniz rule to Eq. (7.31) and using Eqs. (7.58) and (3.20).

Proposition 2 *The scalar product, $(\phi\,|\,\psi)$, Eq. (7.28), is conserved, i.e.,*

$$\frac{d}{dt}\,(\phi\,|\,\psi) = 0, \tag{7.59}$$

where $t = x^0$, and provided ϕ and ψ satisfy the Dirac Eq. (3.30) and its adjoint.

Proposition 2 follows from Proposition 1 provided, as stated in Ref. [1], "we assume that ϕ and ψ, vanish sufficiently rapidly at spatial infinity or obey suitable boundary conditions in a closed universe, so that the spatial components of Eq. (7.31) give vanishing contributions upon integration and the various products are well defined" (c.f. Eq. (7.21)). For more details refer to [1, 5].

7.5 Example: The Closed FRW Universe

In this section we go over the example from Finster and Reintjes, Ref. [4]. We consider the closed FRW universe whose line element, in conformal coordinates, is (this is in lapse and shift form, Eq. (7.29))

$$ds^2 = S(\eta)^2 \left(d\eta^2 - d\chi^2 - f(\chi)^2 \left(d\theta^2 + \sin^2\theta \, d\phi^2\right)\right) . \tag{7.60}$$

In the metric (7.60) η is the conformal time, χ is the radial coordinate, and $\theta \in (0, \pi)$, $\phi \in [0, 2\pi)$, are the angular coordinates. The scale function $S(\eta)$ depends on the type of matter under consideration. In the present example $S(\eta)$ is left unspecified and is an arbitrary positive function. We have

$$f(\chi) = \begin{cases} \sin(\chi), & \text{closed universe,} & \chi \in (0, \pi) \\ \sinh(\chi), & \text{open universe,} & \chi > 0 . \\ \chi, & \text{flat universe,} & \chi > 0 . \end{cases} \tag{7.61}$$

We choose the tetrad vectors in the Cartesian gauge (see Sect. A.3). In this example these tetrad vectors correspond to a class of static observers on the timelike path $\sigma(s) = (\int_0^s \frac{ds}{S(\eta(s))}, \chi_0, \theta_0, \phi_0)$. One can show that $\nabla_{e_0} e_A = 0$, for all A and any well-behaved $S(\eta)$, so the tetrad e_A is a Fermi tetrad field, Eq. (A.15).

$$e_0 = \frac{1}{S(\eta)} \partial_\eta, \tag{7.62}$$

$$e_1 = \frac{\sin\theta \cos\phi}{S(\eta)} \partial_\chi + \frac{\cos\theta \cos\phi}{S(\eta) f(\chi)} \partial_\theta - \frac{\sin\phi}{S(\eta) f(\chi) \sin\theta} \partial_\phi, \tag{7.63}$$

$$e_2 = \frac{\sin\theta \sin\phi}{S(\eta)} \partial_\chi + \frac{\cos\theta \sin\phi}{S(\eta) f(\chi)} \partial_\theta + \frac{\cos\phi}{S(\eta) f(\chi) \sin\theta} \partial_\phi, \tag{7.64}$$

$$e_3 = \frac{\cos\theta}{S(\eta)} \partial_\chi - \frac{\sin\theta}{S(\eta) f(\chi)} \partial_\theta, \tag{7.65}$$

and using Eq. (3.3) we obtain the spacetime-dependent gamma matrices,

$$\bar{\gamma}^\eta = \frac{1}{S(\eta)} \gamma^0, \tag{7.66}$$

$$\bar{\gamma}^\chi = \frac{1}{S(\eta)} \left(\sin\theta \cos\phi \, \gamma^1 + \sin\theta \sin\phi \, \gamma^2 + \cos\theta \, \gamma^3\right), \tag{7.67}$$

$$\bar{\gamma}^\theta = \frac{1}{S(\eta) f(\chi)} \left(\cos\theta \cos\phi \, \gamma^1 + \cos\theta \sin\phi \, \gamma^2 - \sin\theta \, \gamma^3\right), \tag{7.68}$$

$$\bar{\gamma}^\phi = \frac{1}{S(\eta) f(\chi) \sin\theta} \left(-\sin\phi \, \gamma^1 + \cos\phi \, \gamma^2\right). \tag{7.69}$$

In Eqs. (7.66)–(7.69) the γ^A are the constant γ matrices in the standard representation.

We write the Dirac equation as in [4]

$$\left[i\bar{\gamma}^\eta \left(\partial_\eta + \frac{3}{2}\frac{\dot{S}}{S}\right) + i\bar{\gamma}^\chi \left(\partial_\chi + \frac{f'-1}{f}\right) + i\bar{\gamma}^\theta \partial_\theta + i\bar{\gamma}^\phi \partial_\phi - m\right] \Psi = 0, \tag{7.70}$$

where $\dot{S}$ is the derivative with respect to η, and f' is the derivative with respect to χ. Note that $3\dot{S}/(2S) \neq \Gamma_\eta$, etc. What happens here is similar to what happened in deriving Eq. (4.74).

At this point Finster and Reintjes restrict themselves to the closed universe case, $f(\chi) = \sin(\chi)$, and assume a solution of the form

$$\Psi(\eta, \chi, \theta, \phi) = \frac{1}{S(\eta)^{\frac{3}{2}}} \begin{pmatrix} h_1(\eta)\, \psi_\lambda(\chi, \theta, \phi) \\ h_2(\eta)\, \tilde{\psi}_\lambda(\chi, \theta, \phi) \end{pmatrix}. \tag{7.71}$$

We shall examine the probability integral, Eq. (7.27), using the above solution of the Dirac equation in a closed FRW universe. We choose Σ to be a slice of constant conformal time η, then the future-directed normal 1-form n has components (cf. Eq. (7.62)),

$$\left(n_\eta, n_\chi, n_\theta, n_\phi\right) = (S(\eta), 0, 0, 0). \tag{7.72}$$

Using Eqs. (7.27) and (7.66), we have

$$(\Psi \mid \Psi) = \int_\Sigma \bar{\Psi}\, \tilde{\gamma}^\alpha \Psi\, n_\alpha\, d\Sigma, \tag{7.73}$$

$$= \int_\Sigma \bar{\Psi}\, \tilde{\gamma}^\eta \Psi\, n_\eta\, d\Sigma, \tag{7.74}$$

$$= \int_\Sigma \Psi^\dagger \Psi\, d\Sigma. \tag{7.75}$$

In going from Eqs. (7.73) to (7.75), we used Eqs. (7.66) and (7.72). Also

$$d\Sigma = \sqrt{|g_\Sigma|}\, d\chi\, d\theta\, d\phi = S^3(\eta)\, d\mu_{S^3}. \tag{7.76}$$

In Eq. (7.76), g_Σ is the determinant of the induced metric on Σ, and

$$d\mu_{S^3} = \sin^2(\chi) \sin(\theta)\, d\chi\, d\theta\, d\phi, \tag{7.77}$$

is the volume element on the unit sphere S^3 in hyperspherical coordinates. So substituting Eq. (7.71) in Eq. (7.73), we obtain

$$(\Psi \mid \Psi) = \int_\Sigma \Psi^\dagger \Psi\, d\Sigma, \tag{7.78}$$

$$= |h_1|^2 \int_{S^3} |\psi_\lambda|^2 d\mu_{S^3} + |h_2|^2 \int_{S^3} |\tilde{\psi}_\lambda|^2 d\mu_{S^3}, \tag{7.79}$$

$$= |h_1|^2 + |h_2|^2, \tag{7.80}$$

where we have followed the normalization of Ref. [4] (cf. Eq. (D.57)). It follows from Eq. (7.59) that the probability integral is constant in time.

References

1. L. Parker, One-electron atom as a probe of spacetime curvature. Phys. Rev. D **22**, 1922–1934 (1980)
2. L. Parker, D. Toms, *Quantum Field Theory in Curved Spacetime*, (Cambridge U. Press, Cambridge, 2009)
3. S. Carroll, *Spacetime and Geometry, an Introduction to General Relativity*, (Addison-Wesley, San Francisco, 2004)
4. F. Finster, M. Reintjes, The Dirac equation and the normalization of its solutions in a closed Friedmann-Robertson-Walker universe. Class. Quantum Grav. **26**, 105021 (2009). arXiv:0901.0602v4
5. X. Huang, L. Parker, Hermiticity of the Dirac Hamiltonian in curved spacetime. Phys. Rev. D **79**, 024020 (2009)
6. E. Poisson, *A Relativist's Toolkit*, (Cambridge U. Press, Cambridge, 2004)
7. T.C. Chapman, D.J. Leiter, On the generally covariant Dirac equation. Am. J. Phys. **44**, 858–862 (1976)

Appendix A
Tetrads

A.1 The Tetrad Formalism

A tetrad is a set of four linearly independent vectors that can be defined at each point in a (semi-) Riemannian spacetime. Here we give a summary of useful relations for tetrad *fields*. Good detailed discussions can be found in several texts see, for example, Appendix J of [1]. We have the following basic relations that determine the vector fields $e_A{}^\alpha$ or the 1-forms (covector fields) $e^A{}_\alpha$, (we may use the notation, $e^A = e^A{}_\alpha \, dx^\alpha$ and $e_A = e_A{}^\alpha \, \partial_\alpha$). The tetrads by definition satisfy the relations (see [1], Eq. (J.3))

$$e^A{}_\alpha \, e_A{}^\beta = \delta_\alpha{}^\beta \,, \tag{A.1}$$

$$e^A{}_\alpha \, e_B{}^\alpha = \delta^A{}_B \,. \tag{A.2}$$

The choice of the tetrad field determines the metric through Eq. (A.3).

$$g_{\alpha\beta} = e^A{}_\alpha \, e^B{}_\beta \, \eta_{AB} \,, \tag{A.3}$$

$$\eta_{AB} = e_A{}^\alpha \, e_B{}^\beta \, g_{\alpha\beta} \,, \tag{A.4}$$

where η_{AB} is the Minkowski spacetime metric in Cartesian coordinates. We note that through Eq. (A.3) the fields e^A define the spacetime, thus certain authors [2] refer to them as *the gravitational field*. We shall always assume that the velocity vector field, e_0, is tangent to a congruence of timelike paths and thus the tetrads are moving along these paths. The reader should also read the comments in Sect. 3.1 below Eq. (3.3), and in Sect. 3.6 below Eq. (3.100).

Under coordinate transformations, greek indices are treated as tensor indices, while latin indices are merely labels (thus the $e_A{}^\alpha$ represent four different vector fields). Equation (A.4) are also a statement of the orthonormality of the vectors $e_A{}^\alpha$. The tetrad components may be determined using the Eq. (A.3) or (A.4).

© The Author(s), under exclusive license to Springer Nature Switzerland AG 2019
P. Collas and D. Klein, *The Dirac Equation in Curved Spacetime*,
SpringerBriefs in Physics, https://doi.org/10.1007/978-3-030-14825-6

Remark 15 It is easy to convince oneself that relabeling the subscripts (or super-scripts) of the $e_A{}^\alpha$ in a consistent way, does not affect the relation (A.4). However, problems may arise, if one is careless with relabeling and reordering variables while using a symbolic manipulation software.

Although in these notes we have considered spacetimes with dimensionality of two or four, in general, if n is the dimensionality of the manifold, Eq. (A.4) are a set of $(\frac{1}{2})n(n+1)$ equations for the n^2 unknown components of the *vielbein* $e_A{}^\alpha$. Therefore $(\frac{1}{2})n(n-1)$ components can be freely chosen or determined by extra conditions.

Exercise 5 *It is a simple exercise to show that Eqs. (A.2) and (A.3), imply Eq. (A.4), while Eq. (A.1) and (A.4), imply Eq. (A.3).*

We have the following rules for raising and lowering indices,

$$e_{A\alpha} = g_{\alpha\beta} e_A{}^\beta, \tag{A.5}$$

$$e^A{}_\alpha = \eta^{AB} e_{B\alpha}. \tag{A.6}$$

The components of tensors in the tetrad frame are given by relations such as the ones below

$$V^A = e^A{}_\alpha V^\alpha, \tag{A.7}$$

$$T^A{}_B = e^A{}_\alpha e_B{}^\beta T^\alpha{}_\beta, \tag{A.8}$$

and so on. Note that in Eq. (A.7) we are taking the product of a vector V with n 1-forms $e^A{}_\alpha$, as a result, we are replacing the vector V with n scalars V^A. Likewise in Eq. (A.8), we are replacing the $\binom{1}{1}$ tensor T with n^2 scalars $T^A{}_B$ [3].

We can obtain the tensor components in the global chart from the "components" in the tetrad frame using relations like the one below

$$V^\alpha = e_A{}^\alpha V^A = e^{A\alpha} V_A. \tag{A.9}$$

Using the above relations we can show that $U_\mu V^\mu = U_A V^A$.

$$g_{\mu\nu} U^\mu V^\nu = \eta_{AB} e^A{}_\mu e^B{}_\nu U^\mu V^\nu, \tag{A.10}$$

$$g_{\mu\nu} U^\mu V^\nu = \eta_{AB} U^A V^B, \tag{A.11}$$

$$U_\mu V^\mu = U_A V^A. \tag{A.12}$$

A.2   Fermi Tetrad Fields

A Fermi tetrad field must satisfy some special conditions. As usual the tetrad field, satisfies Eq. (A.3), etc., but in the Fermi case, the velocity vector field, e_0, is tangent to a congruence of timelike geodesics, $\sigma(\tau)$, parametrized by the proper time τ. Thus

$$e_0 = \frac{d\sigma(\tau)}{d\tau}, \tag{A.13}$$

and therefore,

$$\nabla_{e_0} e_0 = 0. \tag{A.14}$$

A Fermi tetrad field must satisfy the equations

$$\nabla_{e_0} e_A = 0, \quad A = 0, 1, 2, 3, \tag{A.15}$$

so that all the tetrad vectors are parallelly transported along the chosen congruence of timelike geodesics. (Recall that if $\nabla_u u \neq 0$ but $\nabla_u v = 0$, then v is parallel transported but not on a geodesic).

Remark 16 Given a Fermi tetrad enables one to obtain approximate Fermi coordinates by the well-known process given in [4]. Examples of how to obtain exact Fermi coordinates, in cases where this is possible, were given in [5–8].

A.3   The Cartesian Gauge Tetrad

A tetrad field referred to as the "Cartesian gauge" was introduced by Brill and Wheeler [9], and has been found useful by many authors [10–15]. An example of the Cartesian gauge tetrad used in the FRW universe was given above, Eqs. (7.62)–(7.65). Here we first discuss it in the simplest case namely flat Lorentz spacetime. Consider the standard tetrad vectors in (Cartesian) Minkowski spacetime, namely,

$$e_0 = \partial_t, \qquad\qquad e_1 = \partial_x, \tag{A.16}$$
$$e_2 = \partial_y, \qquad\qquad e_3 = \partial_z.$$

Then, if we transform to spherical coordinates (t, r, θ, ϕ), the above tetrad transforms into the tetrad vector fields below

$$h_0 = \partial_t ,$$

(A.17)

$$h_1 = \sin\theta\cos\phi\,\partial_r + \frac{\cos\theta\cos\phi}{r}\partial_\theta - \frac{\sin\phi}{r\sin\theta}\partial_\phi ,$$

(A.18)

$$h_2 = \sin\theta\sin\phi\,\partial_r + \frac{\cos\theta\sin\phi}{r}\partial_\theta + \frac{\cos\phi}{r\sin\theta}\partial_\phi ,$$

(A.19)

$$h_3 = \cos\theta\,\partial_r - \frac{\sin\theta}{r}\partial_\theta ,$$

(A.20)

For the case of the flat spacetime the Fock-Ivanenko coefficients vanish with this tetrad in spherical coordinates. The metric is the usual spherical coordinate metric,

$$ds^2 = dt^2 - dr^2 - r^2(d\theta^2 + \sin^2\theta\,d\phi^2) .$$

(A.21)

Note that the default tetrad vector fields for the metric of Eq. (A.21), namely,

$$s_0 = \partial_t ,$$

(A.22)

$$s_1 = \partial_r ,$$

(A.23)

$$s_2 = \frac{1}{r}\partial_\theta ,$$

(A.24)

$$s_3 = \frac{1}{r\sin\theta}\partial_\phi ,$$

(A.25)

are "rotated" (see Appendix A.4) with respect to the original Cartesian tetrad Eq. (A.16) and the Fock-Ivanenko coefficients no longer vanish.

Exercise 6 (a) *Write the Dirac equation using the tetrad, (A.17)–(A.20), for the metric (A.21), in the chiral representation Eq. (D.18). (b) Transform the chiral plane wave solution is $\phi^{(+)(1)}$ in Eq. (D.14) to spherical coordinates and show that it satisfies the Dirac equation obtained in part (a).*

A.4 Vielbeins, Spinors and Local Lorentz Transformations

Suppose we solve the Dirac equation in a given manifold using a certain chart and tetrad field. We would like to know how our spinor ψ is related to a spinor ψ' on the same manifold but a different tetrad field related to the first one by a local Lorentz transformation in $\Lambda(x)$. Although we shall prove some of the statements, we refer the reader to the proofs given in [16].

We mention for the sake of clarity that if $F \in G$, where G is a group of coordinate transformations, then we write

$$\bar{x} = Fx .$$

(A.26)

Thus in general for a scalar function, $\phi(x)$, we have

$$\phi(x) = \bar{\phi}(\bar{x}). \tag{A.27}$$

If in a coordinate system (x^0, x^i), we change from an initial chosen vielbein[1] set, h_A, to another set, e_A, then the new vielbein vectors can be expressed as linear combinations of the old,[2]

$$e_A{}^\mu = \Lambda_A{}^B h_B{}^\mu. \tag{A.28}$$

However, both vielbein sets must satisfy Eq. (A.4), i.e.,

$$\eta_{AB} = h_A{}^\alpha h_B{}^\beta g_{\alpha\beta}, \tag{A.29}$$

$$\eta_{AD} = e_A{}^\mu e_D{}^\nu g_{\mu\nu}. \tag{A.30}$$

Substituting Eq. (A.28) in Eq. (A.29), we obtain

$$\Lambda_A{}^B h_B{}^\mu \Lambda_D{}^C h_C{}^\nu g_{\mu\nu} = \eta_{AD}, \tag{A.31}$$

$$\Lambda_A{}^B \Lambda_D{}^C \eta_{BC} = \eta_{AD}, \tag{A.32}$$

$$\Lambda^T \eta \Lambda = \eta, \tag{A.33}$$

where Λ^T is the transpose of Λ. From Eq. (A.33) we have that $\det \Lambda = \pm 1$. Thus it follows then from Eq. (A.32) that $\Lambda_A{}^B$ is a Lorentz matrix. So in the context of general relativity *the Lorentz group is the group of vielbein rotations* [16] p. 143. We also remark that the Λ matrices will in general be spacetime-dependent. We shall refer to them as local Lorentz transformations.

Under coordinate transformations spinors, ψ, behave like scalars so that, [16] p. 147,

$$\begin{pmatrix} \bar{\phi}(\bar{x}) \\ \bar{\chi}(\bar{x}) \end{pmatrix} = \begin{pmatrix} \phi(x) \\ \chi(x) \end{pmatrix}. \tag{A.34}$$

However when a vielbein h_B, is rotated by Λ as in Eq. (A.28), then

$$\psi_e = L\psi_h, \tag{A.35}$$

where L is a (spacetime-dependent) spinor representative of a vielbein rotation Λ, [16] pp. 76, 147,

$$L = \begin{pmatrix} S & 0 \\ 0 & (S^\dagger)^{-1} \end{pmatrix}, \tag{A.36}$$

---

[1]We shall use the term vielbein whenever the dimensionality is not necessarily (3+1). We reserve the term tetrad for the (3+1) case.

[2]A shorthand for Eq. (A.28) is $e = \Lambda^{-1} h$, while Eq. (A.37) is $L^{-1}\gamma L = \Lambda\gamma$, so that our index positions agree with Refs. [17–19].

with $\det(L) = 1$, [17], that satisfies the relations, [16] p. 147,

$$L^{-1}\gamma^A L = \Lambda^A{}_B \gamma^B,$$ (A.37)

and

$$\gamma^0 L^\dagger \gamma^0 = L^{-1}.$$ (A.38)

Given in [20], Eq. (5.396), p. 246. For a derivation of Eq. (A.37) see, e.g., [18].

Below we provide some helpful details for the calculation of the matrix L.

We use the equation given in Ref. [20], p. 225, with appropriate notational changes,

$$L(\Lambda(x)) = \exp\left(i\epsilon_{AB}\Sigma^{AB}\right),$$ (A.39)

where Λ is a local Lorentz transformation, ϵ_{AB} are the parameters characterizing the Lorentz transformation. Also from [20], Eq. (5.284), p. 228, we have that

$$\Sigma^{AB} = -\frac{i}{8}\left[\gamma^A, \gamma^B\right],$$ (A.40)

thus

$$L = \exp\left(\frac{\epsilon_{AB}}{8}\left[\gamma^A, \gamma^B\right]\right).$$ (A.41)

Our γ matrices will be in the chiral representation.

It is straightforward to show that, $\det(L) = 1$. We use the relation that for a matrix M we have that

$$\det\left(e^M\right) = e^{\text{tr}(M)}.$$ (A.42)

Now we use Eq. (A.41) in its most general form. The exponent then is

$$\frac{1}{2}\left(\epsilon_{01}\gamma^0\gamma^1 + \epsilon_{02}\gamma^0\gamma^2 + \epsilon_{03}\gamma^0\gamma^3 + \epsilon_{12}\gamma^1\gamma^2 + \epsilon_{13}\gamma^1\gamma^3 + \epsilon_{23}\gamma^2\gamma^3\right).$$ (A.43)

By evaluating each product of pairs of γ matrices separately in the chiral representation, we find that they all vanish. Thus the trace of expression (A.43) vanishes. This is a representation independent result.

The first step is, of course, to find the local Lorentz transformation Λ involved in the tetrad rotation. This is easily deduced from the relation,

$$e^A{}_\mu = \Lambda^A{}_B h^B{}_\mu.$$ (A.44)

For the details of the relation between Λ and L, namely the calculation of the parameters ϵ_{AB}, a clear presentation is given in Chap. 1 of Hitoshi Yamamoto's lecture notes [21] where it may be seen that Λ may be expressed as

$$\Lambda = e^{\xi_i K_i + \theta_i L_i} \equiv e^M, \qquad i = 1, 2, 3, \tag{A.45}$$

where the matrices K_i and L_i are given in [21], furthermore we find that

$$M = \begin{pmatrix} 0 & \xi_1 & \xi_2 & \xi_3 \\ \xi_1 & 0 & -\theta_3 & \theta_2 \\ \xi_2 & \theta_3 & 0 & -\theta_1 \\ \xi_3 & -\theta_2 & \theta_1 & 0 \end{pmatrix} = \begin{pmatrix} 0 & \epsilon_{01} & \epsilon_{02} & \epsilon_{03} \\ \epsilon_{01} & 0 & -\epsilon_{12} & -\epsilon_{13} \\ \epsilon_{02} & \epsilon_{12} & 0 & -\epsilon_{23} \\ \epsilon_{03} & \epsilon_{13} & \epsilon_{23} & 0 \end{pmatrix}, \tag{A.46}$$

or

$$\epsilon_{01} = \xi_1, \ \epsilon_{02} = \xi_2, \ \epsilon_{03} = \xi_3, \ \epsilon_{12} = \theta_3, \ \epsilon_{13} = -\theta_2, \ \epsilon_{23} = \theta_1. \tag{A.47}$$

If we have a Λ, then using Mathematica's MatrixLog[Λ], we obtain our expression for M. Comparing coefficients with (A.41), we find the ϵ_{AB}. Finally Mathematica's MatrixExp[M] gives us L. Apart from the simple example in our Sect. 5.2 above, a more elaborate calculation can be found in Sect. 6 of Ref. [22].

Appendix B
The Gamma Matrices

B.1 General Summary

The 2×2 Pauli spin matrices are

$$\sigma^1 = \begin{pmatrix} 0 & 1 \\ 1 & 0 \end{pmatrix}, \quad \sigma^2 = \begin{pmatrix} 0 & -i \\ i & 0 \end{pmatrix}, \quad \sigma^3 = \begin{pmatrix} 1 & 0 \\ 0 & -1 \end{pmatrix}. \tag{B.1}$$

For a free spin $1/2$ particle of mass m we write the Dirac equation in Minkowski spacetime as

$$i\gamma^A \partial_A \psi - m\psi = 0, \tag{B.2}$$

where ψ is a 4-component (contravariant) spinor and the 4×4 γ (constant) matrices,[3] satisfy the anticommutation relation

$$\{\gamma^A, \gamma^B\} = \varepsilon\, 2\eta^{AB} I, \tag{B.3}$$

where $\varepsilon = \pm 1$, and the Hermiticity conditions

$$(\gamma^A)^\dagger = \gamma^0 \gamma^A \gamma^0. \tag{B.4}$$

We raise and lower the indices using the metric η, e.g., $\gamma^A = \eta^{AB}\gamma_B$.

All representations below satisfy the relation

$$\{\gamma^A, \gamma^B\} = 2\eta^{AB} I, \tag{B.5}$$

---

[3] So more precisely Eq. (B.2) is $i(\gamma^A)^i{}_k \partial_A \psi^k - m\psi^i = 0$, $i, k = (1, 2, 3, 4)$.

© The Author(s), under exclusive license to Springer Nature Switzerland AG 2019
P. Collas and D. Klein, *The Dirac Equation in Curved Spacetime*,
SpringerBriefs in Physics, https://doi.org/10.1007/978-3-030-14825-6

with signature convention $(+, -, -, -)$. We also define the matrix

$$\gamma^5 := i\,\gamma^0\gamma^1\gamma^2\gamma^3\,, \tag{B.6}$$

which satisfies the representation and signature independent relations,

$$\{\gamma^A, \gamma^5\} = 0\,, \quad \left(\gamma^5\right)^2 = I_4\,, \quad \left(\gamma^5\right)^\dagger = \gamma^5\,. \tag{B.7}$$

Remark 17 The sign choice in Eq. (B.3) depends on the metric sign convention and the representation of the γ matrices. There are several commonly used representations each with its own advantages. One can avoid the $(\varepsilon = -1)$ choice in Eq. (B.3) by multiplying the γ matrices with $\pm i$, (e.g., both [2, 23] multiply by $-i$). These matters are further elucidated in Appendix C.

B.2   The Standard or Dirac-Pauli Representation

In the *standard or Dirac-Pauli representation*, (or the *Bjorken-Drell representation*) we have

$$\gamma^0 = \begin{pmatrix} I_2 & 0 \\ 0 & -I_2 \end{pmatrix}, \quad \gamma^K = \begin{pmatrix} 0 & \sigma^K \\ -\sigma^K & 0 \end{pmatrix}, \quad K = (1, 2, 3)\,. \tag{B.8}$$

It is easy to verify that,

$$\left(\gamma^0\right)^2 = I\,, \quad \left(\gamma^K\right)^2 = -I\,. \tag{B.9}$$

We also give below the Dirac β and α^K matrices,

$$\beta = \begin{pmatrix} I_2 & 0 \\ 0 & -I_2 \end{pmatrix}, \quad \alpha^K = \begin{pmatrix} 0 & \sigma^K \\ \sigma^K & 0 \end{pmatrix}, \quad K = (1, 2, 3)\,, \tag{B.10}$$

that is, $\gamma^0 = \beta$, $\gamma^K = \beta\alpha^K$.

B.3   The Chiral or Weyl Representation

In the *chiral or Weyl representation*, there are two possible choices for γ^0, we choose

$$\gamma^0 = \begin{pmatrix} 0 & -I_2 \\ -I_2 & 0 \end{pmatrix}, \quad \gamma^K = \begin{pmatrix} 0 & \sigma^K \\ -\sigma^K & 0 \end{pmatrix}, \quad K = (1, 2, 3)\,, \tag{B.11}$$

and

$$\left(\gamma^0\right)^2 = I\,, \quad \left(\gamma^K\right)^2 = -I\,. \tag{B.12}$$

Another option, in the chiral representation, is to choose the negative of the above γ^0, in which case γ^5 changes sign, unless one defines it as the negative of Eq. (B.6). Some authors define the chiral γ matrices by multiplying all of the γ's in Eq. (D.18) by (-1), then the γ^5 does not change sign. In any of the above-mentioned chiral representations, the chirality operator γ^5 given by Eq. (B.6) is equal to

$$\gamma^5 = \pm \begin{pmatrix} I_2 & 0 \\ 0 & -I_2 \end{pmatrix}, \tag{B.13}$$

(see also Sect. D.1).

B.4 The Majorana Representation

In the *Majorana representation* the γ matrices are imaginary and the spinors are real.

$$\gamma^0 = \begin{pmatrix} 0 & \sigma^2 \\ \sigma^2 & 0 \end{pmatrix}, \quad \gamma^1 = \begin{pmatrix} i\sigma^3 & 0 \\ 0 & i\sigma^3 \end{pmatrix}, \tag{B.14}$$

$$\gamma^2 = \begin{pmatrix} 0 & -\sigma^2 \\ \sigma^2 & 0 \end{pmatrix}, \quad \gamma^3 = \begin{pmatrix} -i\sigma^1 & 0 \\ 0 & -i\sigma^1 \end{pmatrix}, \tag{B.15}$$

and

$$\left(\gamma^0\right)^2 = I, \quad \left(\gamma^K\right)^2 = -I. \tag{B.16}$$

B.5 The Jauch-Rohrlich Representation

In the *Jauch-Rorhlich representation* [24], we have

$$\gamma^0 = -i \begin{pmatrix} I_2 & 0 \\ 0 & -I_2 \end{pmatrix}, \quad \gamma^K = \begin{pmatrix} 0 & \sigma^K \\ \sigma^K & 0 \end{pmatrix}, \quad K = (1, 2, 3), \tag{B.17}$$

in fact form Eq. (B.10) we have that,

$$\gamma^0 = -i\beta, \quad \gamma^K = \alpha^K, \tag{B.18}$$

thus,

$$\left(\gamma^0\right)^2 = -I, \quad \left(\gamma^K\right)^2 = I, \tag{B.19}$$

and we satisfy

$$\{\gamma^A, \gamma^B\} = 2\eta^{AB} I, \tag{B.20}$$

with signature $(-, +, +, +)$.

Jauch and Rorhlich define $\gamma^5 \equiv \gamma_5 \equiv \gamma^0\gamma^1\gamma^2\gamma^3$, so again γ^5 satisfies Eq. (B.7). Furthermore, instead of Eq. (B.2), we now have

$$\gamma^A \partial_A \psi + m\psi = 0 \, . \tag{B.21}$$

B.6   Gamma Matrices in $(1+1)$ and $(2+1)$ Spacetime

We give a few examples of γ matrix sets in lower spacetime dimensionalities, [20, 25].

In $(1+1)$ a chiral representation set is

$$\gamma^0 = \sigma^1, \quad \gamma^1 = \pm i\sigma^2, \tag{B.22}$$

with $\gamma^5 \equiv \gamma^3 = \sigma^3$.

A Majorana representation is

$$\gamma^0 = \sigma^2, \quad \gamma^1 = i\sigma^1. \tag{B.23}$$

In $(2+1)$ spacetime we have the two standard representations which are not related by any transformation [25],

$$\gamma^0 = \sigma^3, \quad \gamma^1 = i\sigma^2, \quad \gamma^2 = -si\sigma^1, \quad s = \pm 1. \tag{B.24}$$

The chiral representation

$$\gamma^0 = \sigma^1, \quad \gamma^1 = i\sigma^2, \quad \gamma^2 = i\sigma^3, \tag{B.25}$$

and, finally, the Majorana representation

$$\gamma^0 = \sigma^2, \quad \gamma^1 = i\sigma^1, \quad \gamma^2 = -i\sigma^3. \tag{B.26}$$

Appendix C
Metric Signatures, the FI Coefficients, etc.

C.1 Additional Comments on the Gamma Matrices

In this subsection we amplify on a comment made in Ref. [26]. It is clear from Eqs. (3.4), (3.3) and (3.5), that the transformation

$$\gamma^A \to -\gamma^A, \quad \forall \ A, \tag{C.1}$$

changes

$$\bar{\gamma}^\alpha \to -\bar{\gamma}^\alpha, \quad \forall \ \alpha, \tag{C.2}$$

giving rise to another representation of the gamma matrices. We note that the Fock-Ivanenko Γ_μ, do not change since they involve products of two gamma matrices. The solutions, $\psi(-\gamma)$ obtained in the new representation, are related to the $\psi(+\gamma)$ in the old representation by letting $m \to -m$. For example, in the standard representation, we have solved the Dirac equation, $i\gamma^A \partial_A \psi - m\psi = 0$, and obtained the solutions given in Eqs. (2.24)–(2.26), which we denote by the shorthand $\psi(\gamma, m)$. If we now define a new set of γ matrices, which we call δ matrices, such that $\delta^A = -\gamma^A$, where γ^A is the standard represaentation, then the Dirac equation is $i\delta^A \partial_A \psi - m\psi = 0$, and the solutions to this equation are given again by Eqs. (2.24)–(2.26) but with $m \to -m$, i.e., $\psi(\delta, -m)$.

In the Minkowski case we also have the freedom of changing the sign of just one of the γ matrices, e.g., $\gamma^1 \to -\gamma^1$ and leaving the rest unchanged (see comment below Eq. (B.12)). Then for example the new solution is related to the old one by letting $p_x \to -p_x$.

P. Collas and D. Klein, *The Dirac Equation in Curved Spacetime*, SpringerBriefs in Physics, https://doi.org/10.1007/978-3-030-14825-6

C.2   Signature (−2)

For easy reference we begin by recalling our definitions of the spin connection coefficients, $\omega_{AB\mu}$, the spinor affine connection, Γ_μ, the Fock-Ivanenko coefficients, Γ_C, and the anticommutation relations of the γ matrices.

$$\omega_{AB\mu} = g_{\beta\alpha}e_A{}^\alpha \nabla_\mu e_B{}^\beta , \tag{C.3}$$

$$\Gamma_\mu = \frac{\varepsilon}{4}\omega_{AB\mu}\gamma^A\gamma^B , \tag{C.4}$$

$$\Gamma_C = e_C{}^\mu \Gamma_\mu , \tag{C.5}$$

$$\{\gamma^A, \gamma^B\} = \varepsilon\, 2\eta^{AB} I . \tag{C.6}$$

It is clear that the sign of the $\omega_{AB\mu}$ coefficients depends on the signature because of the $g_{\beta\alpha}$ factor in Eq. (C.3). We now let $\varepsilon = +1$ in Eqs. (C.4), (C.6) and use any γ matrix representation whose matrices satisfy

$$\left(\gamma^0\right)^2 = I , \qquad \left(\gamma^K\right)^2 = -I , \tag{C.7}$$

We then obtain Γ_μ and Γ_C, and we may write the Dirac equation as

$$i\gamma^C (e_C + \Gamma_C)\,\psi - m\psi = 0 . \tag{C.8}$$

C.3   Signature (+2)

We begin by letting $\varepsilon = +1$ in Eqs. (C.4), (C.6). The simplest way to satisfy Eq. (C.6) with signature (+2) is to absorb the i in Eq. (C.8) into the definition of a new set of γ matrices so that now, instead of Eq. (C.7), we have

$$\left(\gamma^0\right)^2 = -I , \qquad \left(\gamma^K\right)^2 = I . \tag{C.9}$$

It is easy to see from Eq. (C.4) that the change of signature will also change the sign of the coefficients $\omega_{AB\mu}$. Thus the latter sign change along with the product of the two $(+i)$ factors from the γ matrices in Eq. (C.4), will give us the same Γ_C coefficients as before (Sect. C.2). The Dirac equation is now written as

$$\gamma^C (e_C + \Gamma_C)\,\psi - m\psi = 0 , \tag{C.10}$$

Some authors prefer to multiply their γ matrices with a factor $(-i)$, e.g., [2, 23]. With the definitions in our paper or the ones in Ref. [2] the Dirac equation becomes

$$\gamma^C (e_C + \Gamma_C)\,\psi + m\psi = 0 . \tag{C.11}$$

Parker in [23], using the Dirac β and α^K matrices, Eq. (B.10), has $\gamma^0 = \eta^{00}\gamma_0 = -i\beta$, and $\gamma^K = \gamma^0\alpha^K$. In addition Parker defines his Γ_μ with the opposite sign from the one adopted here and compensates with the usual $(-)$ sign change in the Dirac equation. Thus his Dirac equation is

$$\gamma^C \left(e_C - \Gamma_C\right)\psi + m\psi = 0. \tag{C.12}$$

As another example we consider Ryder in Ref. [27]. Ryder uses $\varepsilon = -1$ in Eqs. (C.4), (C.6), so he can use the usual γ matrix representations with

$$\left(\gamma^0\right)^2 = I, \qquad \left(\gamma^K\right)^2 = -I. \tag{C.13}$$

Note, however, that the $\varepsilon = -1$ along with the sign change due to the signature, ultimately gives the same Γ_μ and Γ_C coefficients as ours obtained in Sect. C.2. Clearly, using $\varepsilon = -1$ is just completely equivalent to multiplying the γ matrices with $(\pm i)$, except that now we don't have to hide the (i) in the Dirac equation, which retains its standard form (see [27], Eq. (11.129)).

$$i\gamma^C \left(e_C + \Gamma_C\right)\psi - m\psi = 0. \tag{C.14}$$

Finally one may use the Jauch-Rohrlich representation. This representation is used, for example, by Hounkonnou and Mendy [28] (see Sects. 4.3 and B.3). In this case the Dirac equation is

$$\gamma^C \left(e_C + \Gamma_C\right)\psi + m\psi = 0. \tag{C.15}$$

Remark 18 It follows from Eqs. (B.7), (C.4)–(C.8), that multiplying the Dirac equation from the left by γ^5 we have

$$i\gamma^5\gamma^C \left(e_C + \Gamma_C\right)\psi - m\gamma^5\psi = 0, \tag{C.16}$$
$$i\gamma^C \left(e_C + \Gamma_C\right)\gamma^5\psi + m\gamma^5\psi = 0. \tag{C.17}$$

Appendix D
Some Further Topics on the Dirac Equation in Special Relativity

D.1 Chiral Representation Set

We begin by stating Pauli's fundamental theorem relating two different representations of the γ matrices in (3+1). spacetime. For some generalizations to other dimensionalities, see [29, 30].

Theorem 2 *Suppose we have two representations of the γ matrices, γ^μ and γ'^μ, satisfying the anticommutation relations*

$$\{\gamma^\mu, \gamma^\nu\} = 2\eta^{\mu\nu} I \,, \tag{D.1}$$

$$\{\gamma'^\mu, \gamma'^\nu\} = 2\eta^{\mu\nu} I \,, \tag{D.2}$$

then there exists a nonsingular 4×4 matrix U, such that,

$$\gamma'^\mu = U\gamma^\mu U^{-1}. \tag{D.3}$$

Moreover if,

$$\left(\gamma^0\right)^\dagger = \gamma^0, \quad \left(\gamma^k\right)^\dagger = -\gamma^k \,, \tag{D.4}$$

$$\left(\gamma'^0\right)^\dagger = \gamma'^0, \quad \left(\gamma'^k\right)^\dagger = -\gamma'^k \,, \quad k = (1, 2, 3), \tag{D.5}$$

then the matrix U can be chosen to be unitary (see [31, 32]).

Exercise 7 *Show that if the relation (D.3) holds, then if $(\gamma^\mu p_\mu - mI)\psi = 0$, then $(\gamma'^\mu p_\mu - mI)\phi = 0$, where $\phi = U\psi$.*

For example, the unitary matrix U below relates the standard representation to our version of the chiral representation, Eq. (D.18).

$$U = \frac{1}{\sqrt{2}} \begin{pmatrix} I_2 & I_2 \\ -I_2 & I_2 \end{pmatrix}. \tag{D.6}$$

© The Author(s), under exclusive license to Springer Nature Switzerland AG 2019
P. Collas and D. Klein, *The Dirac Equation in Curved Spacetime*,
SpringerBriefs in Physics, https://doi.org/10.1007/978-3-030-14825-6

We have that

$$\gamma^{\mu}_{(chiral)} = U\gamma^{\mu}_{(stand)}U^{-1},$$ (D.7)

compare Eqs. (B.8) and (D.18). Using Eq. (2.7) we write

$$iU\gamma^{\mu}U^{-1}U\partial_{\mu}\psi - mU\psi = 0,$$ (D.8)

so instead of Eqs. (2.12)–(2.15), we now have,

$$-i\partial_t\phi_3 + i\partial_z\phi_3 + i\partial_x\phi_4 + \partial_y\phi_4 - m\phi_1 = 0,$$ (D.9)

$$i\partial_x\phi_3 - \partial_y\phi_3 - i\partial_t\phi_4 - i\partial_z\phi_4 - m\phi_2 = 0,$$ (D.10)

$$-i\partial_t\phi_1 - i\partial_z\phi_1 - i\partial_x\phi_2 - \partial_y\phi_2 - m\phi_3 = 0,$$ (D.11)

$$-i\partial_x\phi_1 + \partial_y\phi_1 - i\partial_t\phi_2 + i\partial_z\phi_2 - m\phi_4 = 0,$$ (D.12)

where

$$\phi = U\psi.$$ (D.13)

We use Eq. (D.13) and Eqs. (2.27), (2.28), and obtain the chiral, positive and negative energy, solutions below,

$$\phi^{(+)(1)} = \frac{N}{\sqrt{2}}\begin{pmatrix} 1 - \dfrac{p_z}{p_t+m} \\[2mm] -\dfrac{p_x+ip_y}{p_t+m} \\[2mm] -1 - \dfrac{p_z}{p_t+m} \\[2mm] -\dfrac{p_x+ip_y}{p_t+m} \end{pmatrix} e^{-ip_\mu x^\mu}, \quad \phi^{(+)(2)} = \frac{N}{\sqrt{2}}\begin{pmatrix} \dfrac{-p_x+ip_y}{p_t+m} \\[2mm] 1 + \dfrac{p_z}{p_t+m} \\[2mm] \dfrac{-p_x+ip_y}{p_t+m} \\[2mm] -1 + \dfrac{p_z}{p_t+m} \end{pmatrix} e^{-ip_\mu x^\mu},$$

(D.14)

$$\phi^{(-)(1)} = \frac{N}{\sqrt{2}}\begin{pmatrix} 1 - \dfrac{p_z}{p_t+m} \\[2mm] -\dfrac{p_x+ip_y}{p_t+m} \\[2mm] 1 + \dfrac{p_z}{p_t+m} \\[2mm] \dfrac{p_x+ip_y}{p_t+m} \end{pmatrix} e^{ip_\mu x^\mu}, \quad \phi^{(-)(2)} = \frac{N}{\sqrt{2}}\begin{pmatrix} \dfrac{-p_x+ip_y}{p_t+m} \\[2mm] 1 + \dfrac{p_z}{p_t+m} \\[2mm] \dfrac{p_x-ip_y}{p_t+m} \\[2mm] 1 - \dfrac{p_z}{p_t+m} \end{pmatrix} e^{ip_\mu x^\mu}.$$

The chirality operator γ^5, Eq. (B.6), in the chiral representation is

$$\gamma^5 = \begin{pmatrix} I_2 & 0 \\ 0 & -I_2 \end{pmatrix}. \tag{D.15}$$

We introduce the *bispinor* χ, [33] below to represent the spinors in Eq. (D.14),

$$\chi = \begin{pmatrix} \chi_R \\ \chi_L \end{pmatrix}, \tag{D.16}$$

where each entry is a 2-component spinor. We then have

$$\gamma^5 \chi = \begin{pmatrix} +\chi_R \\ -\chi_L \end{pmatrix}, \tag{D.17}$$

χ_R being right-handed and χ_L left-handed.

D.2 Calculating the Polarizations for $u(p)$

In this section we outline how one may obtain the properly polarized spinors $u(p)$ of Eq. (D.14), cf. [34], pp. 44–48. We adopt the chiral representation of Eq. (D.18),

$$\gamma^0 = \begin{pmatrix} 0 & -I_2 \\ -I_2 & 0 \end{pmatrix}, \quad \gamma^K = \begin{pmatrix} 0 & \sigma^K \\ -\sigma^K & 0 \end{pmatrix}, \quad K = (1, 2, 3), \tag{D.18}$$

where the Pauli spin matrices are as usual

$$\sigma^1 = \begin{pmatrix} 0 & 1 \\ 1 & 0 \end{pmatrix}, \quad \sigma^2 = \begin{pmatrix} 0 & -i \\ i & 0 \end{pmatrix}, \quad \sigma^3 = \begin{pmatrix} 1 & 0 \\ 0 & -1 \end{pmatrix}. \tag{D.19}$$

For a free, spin $1/2$ particle of mass m in Minkowski spacetime, we write the Dirac equation for the momentum-dependent spinor part, $u(p)$, as

$$(\not{p} - mI)u(p) = 0. \tag{D.20}$$

It is easy to show that the general solution of Eq. (D.20) in the rest frame of the particle, where $p = (m, 0, 0, 0)$, is

$$u(p) = N(p) \begin{pmatrix} \xi \\ -\xi \end{pmatrix}, \quad \xi = \begin{pmatrix} a_1 \\ a_2 \end{pmatrix}. \tag{D.21}$$

where $N(p)$ is a normalization factor and ξ is a *constant* 2-component spinor. Conventionally one normalizes ξ so that

$$\xi^\dagger \xi = |a_1|^2 + |a_2|^2 = 1. \tag{D.22}$$

A useful reference for this material is Steane [33] see Fig. E.2. To proceed further we also need to define the *vectors* with matrix components as follows,

$$\sigma = \left(1, \sigma^1, \sigma^2, \sigma^3\right), \tag{D.23}$$

$$\bar\sigma = \left(1, -\sigma^1, -\sigma^2, -\sigma^3\right). \tag{D.24}$$

We use the notation below and we write

$$p \cdot \sigma = \eta_{\mu\nu} p^\mu \sigma^\nu = p_t \sigma^0 - p^x \sigma^1 - p^y \sigma^2 - p^z \sigma^3, \tag{D.25}$$

$$= p_t \sigma^0 + p_x \sigma^1 + p_y \sigma^2 + p_z \sigma^3. \tag{D.26}$$

If we now apply a boost in an arbitrary p direction and operate on the rest frame $u(p)$ with the spinor representation of a finite Lorentz transformation, Λ, the result may be written in a compact from as (see [34], pp. 44–48).

$$u(p) = N(p) \begin{pmatrix} \xi \\ -\xi \end{pmatrix} \xrightarrow{\Lambda} \begin{pmatrix} \sqrt{-p \cdot \sigma}\,(+\xi) \\ \sqrt{-p \cdot \bar\sigma}\,(-\xi) \end{pmatrix}, \tag{D.27}$$

where the minus signs in the square roots are necessary (in our case) in order to obtain real (not imaginary) results. Using Eq. (2.50) we verify that for polarization in the positive z direction

$$\xi = \begin{pmatrix} 1 \\ 0 \end{pmatrix}. \tag{D.28}$$

We now restrict ourselves to a boost in the positive z-direction and we obtain, apart from an overall normalization factor, $N(p)$,

$$u(p) = \begin{pmatrix} \sqrt{p_t - p_z} \\ 0 \\ -\sqrt{p_t + p_z} \\ 0 \end{pmatrix}. \tag{D.29}$$

We can re-write the above solution as,

$$u(p) = \sqrt{\frac{p_t + m}{2}} \begin{pmatrix} 1 - \frac{p_z}{p_t + m} \\ 0 \\ -\left(1 + \frac{p_z}{p_t + m}\right) \\ 0 \end{pmatrix}, \tag{D.30}$$

where we can still change the overall normalization $N(p)$. It is easy to reconcile Eqs. (D.29) and (D.30). Recall that for simplicity we let $p_x = p_y = 0$. Thus

$$p_t^2 - m^2 = p_z^2 = (p^z)^2. \tag{D.31}$$

However we are considering a boost in the *positive z* direction, in which case we must let

$$p^z = \sqrt{p_t^2 - m^2} \geq 0, \tag{D.32}$$

and, with our signature, $p_z \leq 0$. Thus

$$\frac{1}{\sqrt{2}} \left(\sqrt{p_t + m} + \sqrt{p_t - m} \right) = \sqrt{p_t - p_z}, \tag{D.33}$$

$$= \sqrt{\frac{p_t + m}{2}} \left(1 - \frac{p_z}{p_t + m} \right), \tag{D.34}$$

$$\frac{1}{\sqrt{2}} \left(\sqrt{p_t - m} - \sqrt{p_t + m} \right) = -\sqrt{p_t + p_z}, \tag{D.35}$$

$$= \sqrt{\frac{p_t + m}{2}} \left(-1 - \frac{p_z}{p_t + m} \right). \tag{D.36}$$

Obtaining spinors polarized in the x- and y-directions is straightforward using the proper ξ's from Steane's Fig. E.2 [33]. Thus for particles with spin in the positive x-direction we have that

$$\xi = \frac{1}{\sqrt{2}} \begin{pmatrix} 1 \\ 1 \end{pmatrix}, \tag{D.37}$$

while for particles with spin in the positive y-direction we have

$$\xi = \frac{1}{\sqrt{2}} \begin{pmatrix} 1 \\ i \end{pmatrix}. \tag{D.38}$$

In any case one should go back to Eq. (D.27) and do a boost in a general direction involving p_x, p_y, p_z.

D.3 Normalization of ψ

In this subsection we mainly discuss how to determine the normalization constants for the plane wave spinors in the standard representation. The calculations are similar in any γ-matrix representation. For clarity we are going to proceed in two steps and thus obtain two normalization constants, N and $\mathbb{N}$. The normalization factor N is the one appearing in Eqs. (2.24) and (2.25), while the normalization factor $\mathbb{N}$ is required when we integrate over the coordinates x^k in the scalar product Eq. (7.23). Using the notation of Eqs. (2.16) and (2.17), we follow Itzykson and Zuber [35] and adopt the Lorentz invariant normalizations,

$$\bar{u}^{(\alpha)}(p)u^{(\beta)}(p) = \delta^{\alpha\beta}, \qquad\qquad \bar{u}^{(\alpha)}(p)v^{(\beta)}(p) = 0, \qquad (D.39)$$

$$\bar{v}^{(\alpha)}(p)v^{(\beta)}(p) = -\delta^{\alpha\beta}, \qquad\qquad \bar{v}^{(\alpha)}(p)u^{(\beta)}(p) = 0. \qquad (D.40)$$

Remark 19 Different authors adopt different normalizations for the Lorentz invariant product $\bar{\psi}\psi$.

The normalization factor N for the plane waves is easily obtained from Eqs. (D.39), (D.40), using the solutions (2.24) and (2.25). Apart from a possible phase factor, we find that

$$N = \sqrt{\frac{m + p_t}{2m}}. \qquad (D.41)$$

So in the notation of Eqs. (2.27) and (2.28), we have

$$\bar{\psi}^{(+)(\alpha)}(x)\psi^{(+)(\beta)}(x) = \delta^{\alpha\beta}, \qquad (D.42)$$

$$\bar{\psi}^{(-)(\alpha)}(x)\psi^{(-)(\beta)}(x) = -\delta^{\alpha\beta}. \qquad (D.43)$$

Before proceeding to the scalar product Eq. (7.23) we prove certain useful relations. For easy reference for the proofs to follow, we write again some of the formulas derived above:

$$\left(\not{p} - mI\right)u^{(\alpha)}(p) = 0, \quad \bar{u}^{(\alpha)}(p)\left(\not{p} - mI\right) = 0, \qquad (D.44)$$

$$\left(\not{p} + mI\right)v^{(\alpha)}(p) = 0, \quad \bar{v}^{(\alpha)}(p)\left(\not{p} + mI\right) = 0. \qquad (D.45)$$

We shall derive an expression for $j^\mu = \bar{\psi}\gamma^\mu\psi$. In the derivation we make use of Eqs. (D.39)–(D.45). For the positive energy solutions we have,

$$\bar{\psi}^{(+)(\alpha)}\gamma^\mu\psi^{(+)(\beta)} = \bar{u}^{(\alpha)}(p)\gamma^\mu u^{(\beta)}(p),$$

$$= \frac{1}{2}\left[\left(\bar{u}^{(\alpha)}\gamma^\mu\right)u^{(\beta)} + \bar{u}^{(\alpha)}\left(\gamma^\mu u^{(\beta)}\right)\right],$$

$$= \frac{1}{2m}\left[\left(\bar{u}^{(\alpha)}m\gamma^\mu\right)u^{(\beta)} + \bar{u}^{(\alpha)}\left(\gamma^\mu m u^{(\beta)}\right)\right],$$

$$= \frac{1}{2m}\left[\left(\bar{u}^{(\alpha)}\not{p}\gamma^\mu\right)u^{(\beta)} + \bar{u}^{(\alpha)}\left(\gamma^\mu \not{p} u^{(\beta)}\right)\right],$$

$$= \frac{1}{2m}\left[\bar{u}^{(\alpha)}\{\not{p},\gamma^\mu\}u^{(\beta)}\right],$$

$$= \frac{1}{2m}\left[\bar{u}^{(\alpha)}p_\nu\{\gamma^\nu,\gamma^\mu\}u^{(\beta)}\right],$$

$$= \frac{1}{2m}\left[\bar{u}^{(\alpha)}p_\nu 2\eta^{\nu\mu}Iu^{(\beta)}\right] = \frac{p^\mu}{m}\delta^{\alpha\beta}. \tag{D.46}$$

Repeating this derivation for the negative energy solutions we get

$$\bar{\psi}^{(-)(\alpha)}\gamma^\mu\psi^{(-)(\beta)} = \bar{v}^{(\alpha)}(p)\gamma^\mu v^{(\beta)}(p),$$

$$= -\frac{1}{2m}\left[\bar{v}^{(\alpha)}\{\not{p},\gamma^\mu\}v^{(\beta)}\right],$$

$$= -\frac{1}{2m}\left[\bar{v}^{(\alpha)}p_\nu 2\eta^{\nu\mu}Iv^{(\beta)}\right] = \frac{p^\mu}{m}\delta^{\alpha\beta}. \tag{D.47}$$

It is important to show that positive and negative energy states are mutually orthogonal if we consider states with opposite energies *but the same* 3-*momentum*. We use the *vector momenta* (see Remark 2),

$$p = (p^t, \boldsymbol{p}), \tag{D.48}$$

$$q = (p^t, -\boldsymbol{p}), \tag{D.49}$$

and write explicitly

$$\psi^{(+)(\alpha)}(x) = u^{(\alpha)}(p)e^{-i\left(p^t t - p^i x^i\right)}, \tag{D.50}$$

$$\psi^{(-)(\beta)}(x) = v^{(\beta)}(q)e^{i\left(p^t t + p^i x^i\right)}. \tag{D.51}$$

Therefore, using again Eqs. (D.39)–(D.45), we have

$$\bar{\psi}^{(-)(\beta)}\gamma^0\psi^{(+)(\alpha)} = e^{-2ip^t t}\bar{v}^{(\beta)}(q)\gamma^0 u^{(\alpha)}(p),$$

$$= \frac{1}{2m}e^{-2ip^t t}\left[\left(\bar{v}^{(\beta)}(q)m\right)\gamma^0 u^{(\alpha)}(p) + \bar{v}^{(\beta)}(q)\gamma^0\left(mu^{(\alpha)}(p)\right)\right],$$

$$= \frac{1}{2m}e^{-2ip^t t}\bar{v}^{(\beta)}(q)\left(-\not{q}\gamma^0 + \gamma^0\not{p}\right)u^{(\alpha)}(p) = 0. \tag{D.52}$$

Showing the last step above requires care!

Exercise 8 *Carry out the details of the last line of Eq. (D.52).*

At this point we introduce the normalization factor, $\mathbb{N}$, which is required by the integration over space in the scalar product below. Thus we write

$$\Psi^{(+)(\alpha)}(x) = \mathbb{N}\psi^{(+)(\alpha)}(x) = \mathbb{N}u^{(\alpha)}(p)e^{-ip_\mu x^\mu}, \tag{D.53}$$

$$\Psi^{(-)(\alpha)}(x) = \mathbb{N}\psi^{(-)(\alpha)}(x) = \mathbb{N}v^{(\alpha)}(p)e^{ip_\mu x^\mu}. \tag{D.54}$$

We now define the scalar product with the standard delta function normalization for free particles. Using Eqs. (D.53), (D.54), we have

$$\left(\Psi_{p'}^{(\alpha)}|\Psi_p^{(\beta)}\right) = \mathbb{N}^2 \int \bar{\psi}_{p'}^{(\pm)(\alpha)}\gamma^0\,\psi_p^{(\pm)(\beta)}d^3x = \mathbb{N}^2\frac{p^t}{m}\,\delta^{\alpha\beta}(2\pi)^3\delta^3\left(\boldsymbol{p}-\boldsymbol{p'}\right), \tag{D.55}$$

where the delta function is given by

$$\delta^3\left(\boldsymbol{p}-\boldsymbol{p'}\right) = \frac{1}{(2\pi)^3}\int e^{i(\boldsymbol{p}-\boldsymbol{p'})\boldsymbol{x}}d^3x, \quad \boldsymbol{x}=\left(x^1,x^2,x^3\right)=(x,y,z). \tag{D.56}$$

The notation in Eq. (D.55) is further elucidated in the solution of Exercise 9.

Exercise 9 *Carry out the details of Eq. (D.55) for the two positive energy spinors.*

The normalization constant $\mathbb{N}$, is chosen so that

$$\left(\Psi_{p'}^{(\alpha)}|\Psi_p^{(\beta)}\right) = \delta^{\alpha\beta}\delta^3\left(\boldsymbol{p}-\boldsymbol{p'}\right). \tag{D.57}$$

Thus

$$\mathbb{N} = \frac{1}{(2\pi)^{\frac{3}{2}}}\sqrt{\frac{m}{p^t}}. \tag{D.58}$$

As a consequence of result (D.52) we see that if, in the integrand of (D.55), the ψ's have opposite energies, the result of the integration is zero. Thus we have that

$$\Psi^{(+)(\alpha)}(x) = \frac{1}{(2\pi)^{\frac{3}{2}}}\sqrt{\frac{m}{p_t}}\,u^{(\alpha)}(p)e^{-ip_\mu x^\mu}, \tag{D.59}$$

$$\Psi^{(-)(\alpha)}(x) = \frac{1}{(2\pi)^{\frac{3}{2}}}\sqrt{\frac{m}{p_t}}\,v^{(\alpha)}(p)e^{ip_\mu x^\mu}. \tag{D.60}$$

Appendix E
Examples of Using CARTAN in General Relativity

Since the software includes its own manual, we are not going to re-write it here. In this Appendix we will briefly explain how one may use the software package CARTAN [36] to obtain the necessary quantities in order to write the Dirac equation in Sects. 4.1 and 4.2. The software comes with a set of 'library' text files for various (3+1) spacetimes, however, one will undoubtedly wish to write and add library files for spacetimes of interest.

E.1 Using CARTAN in the F-I Approach

In this subsection we show how to use the software package CARTAN to obtain the results of Sect. 4.1. This software can calculate the FI coefficients and the expression $i\gamma^C (e_C - \Gamma_C)\,\psi$ as outlined below. We reproduce in the figures below the relevant pages from the CARTAN files and comment on them

CARTAN defines the Γ_C coefficients with the opposite sign from ours. This is compensated in CARTAN by inserting another minus sign so that the Dirac equation now is

$$i\gamma^C (e_C - \Gamma_C)\,\psi - m\psi = 0\,, \tag{E.1}$$

thus ultimately identical to our Eq. (C.8), (Figs. E.1, E.3, E.4, E.5, E.6, E.7, E.8, E.9, E.11, E.12 and E.13).

Furthermore, if one wishes to use signature $(+2)$, then CARTAN multiplies the γ matrices with $(+i)$, so that the Dirac equation is now

$$\gamma^C (e_C - \Gamma_C)\,\psi - m\psi = 0\,. \tag{E.2}$$

© The Author(s), under exclusive license to Springer Nature Switzerland AG 2019
P. Collas and D. Klein, *The Dirac Equation in Curved Spacetime*,
SpringerBriefs in Physics, https://doi.org/10.1007/978-3-030-14825-6

```
(* $Id: Schwarzschild.lib

|


(* **************************************************************** *)
(*                                                                  *)
(*              CARTAN library of geometries                        *)
(*                    (c) H. H. Soleng                              *)
(*                                                                  *)
(*                 Schwarzschild geometry                           *)
(*                 ----------------------                           *)
(* **************************************************************** *)



(*    Orthonormal basis:                                          *)


nDim = 4


Coordinates= If[$VersionNumber>=3,
                {r, \[Theta], \[Phi], t},

                {r, theta, phi, t}]

OrientationOfFrame = -1

Tetrad =   If[$VersionNumber>=3,
   {{(1 - (2*M)/r)^(-1/2), 0, 0, 0}, {0, r, 0, 0},
   {0, 0, r*Sin[\[Theta]], 0}, {0, 0, 0, (1 - (2*M)/r)^(1/2)}},

   {{(1 - (2*M)/r)^(-1/2), 0, 0, 0}, {0, r, 0, 0},
   {0, 0, r*Sin[theta], 0}, {0, 0, 0, (1 - (2*M)/r)^(1/2)}}]

FlatMetric = {{{1, 0, 0, 0}, {0, 1, 0, 0}, {0, 0, 1, 0}, {0, 0, 0, -1}},
   {1, 1}}

Riemannian = True

Torsion = "The manifold is Riemannian."
```

Fig. E.1 This is a slightly edited CARTAN library text file for the Schwarzschild spacetime. Indices are always in the order (1, 2, 3, 4) where 4 stands for the timelike coordinate if there is one. Note the duplication of the coordinates and tetrad fields with the important small differences. The FlatMetric determines the signature

> Here we calculate the Dirac equation in Schwarzschild spacetime with signature (+2). Although we use Ryder's "Introduction to General Relativity" conventions, we do calculate the Fock-Ivanenko coefficients using our Eq. (3.26) with $\varepsilon = -1$.

In[ ]:=
```
Unprotect[RiemannR]
Remove[RiemannR]
Unprotect[Symmetrize]
Remove[Symmetrize]
```

Out[ ]=
```
{RiemannR}
```

Out[ ]=
```
{Symmetrize}
```

> The input file: Schwarzschild.lib is included in CARTAN's library. After executing the next command, type 2 for input from file, then copy and paste Schwarzschild.lib, then type y as the 'yes' answer to the next three queries. CARTAN produces the *structure coefficients*, the *connection*, and the *metric*.

> Input File : Schwarzschild(+2).lib

In[ ]:=
```
<< cartan.m
```

```
- - - - - - - - - - - - - - - - - - - - - - - - - - - - - - - -
     Tensors in Physics: CARTAN 1.8
     by Harald H. Soleng, December 28, 2011
     Copyright © 2002-2011 by Ad Infinitum
     http://www.adinfinitum.no/cartan
- - - - - - - - - - - - - - - - - - - - - - - - - - - - - - - -
```

```
Geometry given by:  1) interactive input

                    2) input from file
```

```
Selection [1]: 2
```

Fig. E.2 This is the page 1 of the CARTAN evaluation of the F-I coefficients. Indices are always in the order (1, 2, 3, 4). In using CARTAN with Mathematica 11, we found it useful to execute the preamble below the first gray cell above. Executing the command, ≪cartan.m, opens a window in which one has to type the answers to certain queries (see the 2nd gray cell above)

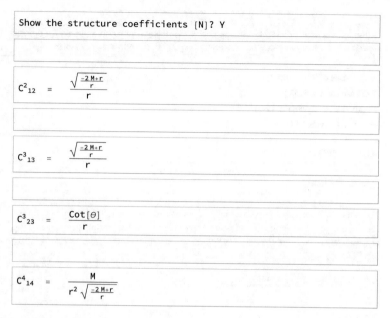

Fig. E.3 This is page 2 of the CARTAN evaluation of the F-I coefficients. The *structure coefficients* above are the same as the *structure constants*

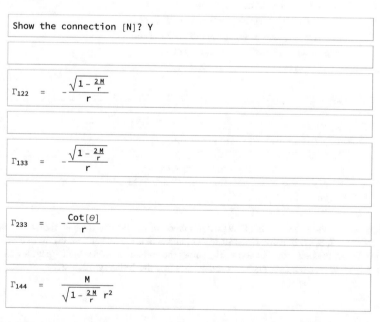

Fig. E.4 This is page 3 of the CARTAN evaluation of the F-I coefficients. The *connection* above, are the *Ricci rotation coefficients*. The Γ_{ABC} are the negatives of ours because CARTAN's C_{ABC} are the negatives of ours in the text

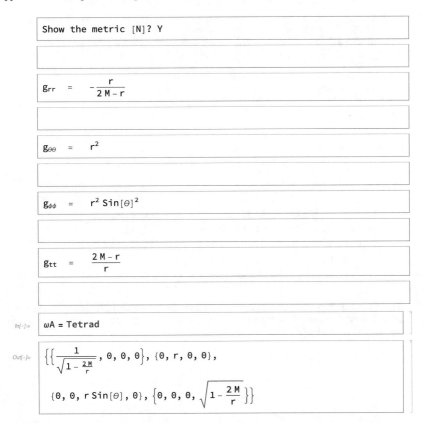

```
Show the metric [N]? Y
```

$$g_{rr} = -\frac{r}{2M - r}$$

$$g_{\theta\theta} = r^2$$

$$g_{\phi\phi} = r^2 \, Sin[\theta]^2$$

$$g_{tt} = \frac{2M - r}{r}$$

In[ ]:= `ωA = Tetrad`

Out[ ]= $\left\{\left\{\dfrac{1}{\sqrt{1 - \dfrac{2M}{r}}}, 0, 0, 0\right\}, \{0, r, 0, 0\}, \right.$

$\left. \{0, 0, r \, Sin[\theta], 0\}, \left\{0, 0, 0, \sqrt{1 - \dfrac{2M}{r}}\right\}\right\}$

Fig. E.5 This is page 4 of the CARTAN evaluation of the F-I coefficients. We have the metric and the input 1-form fields which we have called ωA. One has to ask for them if one is checking for possible input errors

At this point we choose the γ matrix representation. The "D" below is for the Standard "Dirac-Pauli" representation. From top to bottom below they are $\gamma^1, \gamma^2, \gamma^3, \gamma^4, \gamma^5$, where $\gamma^4 = \gamma^0$, and $\gamma^5 = \gamma^4 \gamma^1 \gamma^2 \gamma^3$.

In[ ]:= `Spinor["D"]`

`Representation`

Out[ ]= `Dirac-Pauli`

In[ ]:= `GammaMatrix`

Out[ ]= `{{{0, 0, 0, i}, {0, 0, i, 0}, {0, -i, 0, 0}, {-i, 0, 0, 0}},`
`{{0, 0, 0, 1}, {0, 0, -1, 0}, {0, -1, 0, 0}, {1, 0, 0, 0}},`
`{{0, 0, i, 0}, {0, 0, 0, -i}, {-i, 0, 0, 0}, {0, i, 0, 0}},`
`{{i, 0, 0, 0}, {0, i, 0, 0}, {0, 0, -i, 0}, {0, 0, 0, -i}},`
`{{0, 0, -i, 0}, {0, 0, 0, -i}, {-i, 0, 0, 0}, {0, -i, 0, 0}}}`

Fig. E.6 This is page 5 of the CARTAN evaluation of the F-I coefficients. We have chosen the standard Dirac-Pauli representation. Note that the γ's (given in the order 1, 2, 3, 4, 5) include a factor of i but $\gamma^5 = \gamma^4 \gamma^1 \gamma^2 \gamma^3$

The Cartan FI coefficients are given by

$$\Gamma_C = -\tfrac{1}{4}\Gamma_{ABC}\,\gamma^A\gamma^B$$

which is formally equal to ours, but they differ by a (-) sign from our equation.

In[ ]:= `FockIvanenko`

Out[ ]=
$$\Big\{\{\{0, 0, 0, 0\}, \{0, 0, 0, 0\}, \{0, 0, 0, 0\}, \{0, 0, 0, 0\}\},$$

$$\Big\{\Big\{\frac{i\sqrt{1-\frac{2M}{r}}}{2\,r}, 0, 0, 0\Big\}, \Big\{0, -\frac{i\sqrt{1-\frac{2M}{r}}}{2\,r}, 0, 0\Big\}, \Big\{0, 0, \frac{i\sqrt{1-\frac{2M}{r}}}{2\,r}, 0\Big\},$$

$$\Big\{0, 0, 0, -\frac{i\sqrt{1-\frac{2M}{r}}}{2\,r}\Big\}\Big\}, \Big\{\Big\{0, \frac{1}{4}\,i\left(\frac{2\,i\sqrt{1-\frac{2M}{r}}}{r} + \frac{2\,\mathrm{Cot}[\theta]}{r}\right), 0, 0\Big\},$$

$$\Big\{\frac{1}{4}\,i\left(-\frac{2\,i\sqrt{1-\frac{2M}{r}}}{r} + \frac{2\,\mathrm{Cot}[\theta]}{r}\right), 0, 0, 0\Big\},$$

$$\Big\{0, 0, 0, \frac{1}{4}\,i\left(\frac{2\,i\sqrt{1-\frac{2M}{r}}}{r} + \frac{2\,\mathrm{Cot}[\theta]}{r}\right)\Big\},$$

$$\Big\{0, 0, \frac{1}{4}\,i\left(-\frac{2\,i\sqrt{1-\frac{2M}{r}}}{r} + \frac{2\,\mathrm{Cot}[\theta]}{r}\right), 0\Big\}\Big\},$$

$$\Big\{\Big\{0, 0, 0, -\frac{M}{2\sqrt{1-\frac{2M}{r}}\;r^2}\Big\}, \Big\{0, 0, -\frac{M}{2\sqrt{1-\frac{2M}{r}}\;r^2}, 0\Big\},$$

$$\Big\{0, -\frac{M}{2\sqrt{1-\frac{2M}{r}}\;r^2}, 0, 0\Big\}, \Big\{-\frac{M}{2\sqrt{1-\frac{2M}{r}}\;r^2}, 0, 0, 0\Big\}\Big\}\Big\}$$

In the above FI we have:

$\{\{\{\Gamma_1\}\}, \{\{\Gamma_2\}\}, \{\{\Gamma_3\}\}, \{\{\Gamma_4\}\}\}$, i.e., the first row above is Γ_1.

Fig. E.7 This is page 6 of the CARTAN evaluation of the F-I coefficients. CARTAN gives the F-I coefficients as a set of matrices Γ_C. They differ from ours by a sign which is compensated for later in the calculation

In[·]:= $\psi = \{\psi1[r, \theta, \phi, t], \psi2[r, \theta, \phi, t], \psi3[r, \theta, \phi, t], \psi4[r, \theta, \phi, t]\};$

The equations below are, with Cartan's conventions,
$\gamma^\mu \nabla_\mu \psi = \gamma^\mu (I \partial_\mu - \Gamma_\mu) \psi = \gamma^A (I e_A{}^\mu \partial_\mu - \Gamma_A) \psi$.

In[·]:= `FullSimplify[SlashCDerivative[ψ]]`

Out[·]:= $\left\{ \frac{1}{2\sqrt{1-\frac{2M}{r}}\,r^2} \left(\left(-3\,i\,M + 2\,i\,r + \sqrt{1-\frac{2M}{r}}\,r\,\mathrm{Cot}[\theta] \right) \psi4[r, \theta, \phi, t] + \right. \right.$

$2\,r \left(i\,r\,\psi1^{(0,0,0,1)}[r, \theta, \phi, t] + \sqrt{1-\frac{2M}{r}} \left(i\,\mathrm{Csc}[\theta]\,\psi3^{(0,0,1,0)}[r, \theta, \phi, t] + \psi4^{(0,1,0,0)}[r, \theta, \phi, t] \right) - \right.$

$\left. \left. \left. i\,(2M-r)\,\psi4^{(1,0,0,0)}[r, \theta, \phi, t] \right) \right), \frac{1}{2\sqrt{1-\frac{2M}{r}}\,r^2} \right.$

$\left. \left(\left(-3\,i\,M + 2\,i\,r - \sqrt{1-\frac{2M}{r}}\,r\,\mathrm{Cot}[\theta] \right) \psi3[r, \theta, \phi, t] + 2\,r \left(i\,r\,\psi2^{(0,0,0,1)}[r, \theta, \phi, t] + \sqrt{1-\frac{2M}{r}} \right. \right. \right.$

$\left. \left. \left. \left(-i\,\mathrm{Csc}[\theta]\,\psi4^{(0,0,1,0)}[r, \theta, \phi, t] - \psi3^{(0,1,0,0)}[r, \theta, \phi, t] \right) - i\,(2M-r)\,\psi3^{(1,0,0,0)}[r, \theta, \phi, t] \right) \right),$

$\frac{1}{2\sqrt{1-\frac{2M}{r}}\,r^2} \left(\left(3\,i\,M - 2\,i\,r - \sqrt{1-\frac{2M}{r}}\,r\,\mathrm{Cot}[\theta] \right) \psi2[r, \theta, \phi, t] - 2\,i\,r \left(r\,\psi3^{(0,0,0,1)}[r, \theta, \phi, t] + \sqrt{1-\frac{2M}{r}} \right. \right.$

$\left. \left. \left(\mathrm{Csc}[\theta]\,\psi1^{(0,0,1,0)}[r, \theta, \phi, t] - i\,\psi2^{(0,1,0,0)}[r, \theta, \phi, t] \right) + (-2M+r)\,\psi2^{(1,0,0,0)}[r, \theta, \phi, t] \right) \right),$

$\frac{1}{2\sqrt{1-\frac{2M}{r}}\,r^2} \left(\left(3\,i\,M - 2\,i\,r + \sqrt{1-\frac{2M}{r}}\,r\,\mathrm{Cot}[\theta] \right) \psi1[r, \theta, \phi, t] + 2\,r \left(-i\,r\,\psi4^{(0,0,0,1)}[r, \theta, \phi, t] + \sqrt{1-\frac{2M}{r}} \right. \right.$

$\left. \left. \left. \left(i\,\mathrm{Csc}[\theta]\,\psi2^{(0,0,1,0)}[r, \theta, \phi, t] + \psi1^{(0,1,0,0)}[r, \theta, \phi, t] \right) + i\,(2M-r)\,\psi1^{(1,0,0,0)}[r, \theta, \phi, t] \right) \right) \right\}$

Fig. E.8 This is page 7 of the CARTAN evaluation of the F-I coefficients. We define our ψ as an array with four components. SlashDerivative[ψ] $= \gamma^\mu \left(I \partial_\mu - \Gamma_\mu \right)$, see gray cell above. CARTAN's Γ_μ is the negative of ours hence the $-$ sign

Recalling that the γ matrices already have an i factor in them, we write the Dirac equation, with Cartan's conventions, as
$\gamma^\mu \nabla_\mu \psi - m\psi = \gamma^\mu (I \partial_\mu - \Gamma_\mu) \psi - m\psi = \gamma^A (I e_A{}^\mu \partial_\mu - \Gamma_A) \psi - m\psi = 0$.

`In[ ]:=`   `FullSimplify[SlashCDerivative[ψ] - m ψ]`

`Out[ ]=` $\Big\{ \dfrac{1}{2\sqrt{1-\frac{2M}{r}}\, r^2} \Big(-3\, i\, M\, \psi 4[r, \theta, \phi, t] +$

$r\Big(-2\, m\, \sqrt{1-\dfrac{2M}{r}}\, r\, \psi 1[r, \theta, \phi, t] + \sqrt{1-\dfrac{2M}{r}}\, \big(\mathrm{Cot}[\theta]\, \psi 4[r, \theta, \phi, t] + 2\, i\, \mathrm{Csc}[\theta]\, \psi 3^{(0,0,1,0)}[r, \theta, \phi, t] +$

$2\, \psi 4^{(0,1,0,0)}[r, \theta, \phi, t]\big) + 2\, i\, r\, \big(\psi 1^{(0,0,0,1)}[r, \theta, \phi, t] + \psi 4^{(1,0,0,0)}[r, \theta, \phi, t]\big) +$

$2\, i\, \big(\psi 4[r, \theta, \phi, t] - 2\, M\, \psi 4^{(1,0,0,0)}[r, \theta, \phi, t]\big) \Big) \Big),$

$\dfrac{1}{2\sqrt{1-\frac{2M}{r}}\, r^2} \Big(-3\, i\, M\, \psi 3[r, \theta, \phi, t] - r\Big(2\, m\, \sqrt{1-\dfrac{2M}{r}}\, r\, \psi 2[r, \theta, \phi, t] +$

$\sqrt{1-\dfrac{2M}{r}}\, \big(\mathrm{Cot}[\theta]\, \psi 3[r, \theta, \phi, t] + 2\, i\, \mathrm{Csc}[\theta]\, \psi 4^{(0,0,1,0)}[r, \theta, \phi, t] + 2\, \psi 3^{(0,1,0,0)}[r, \theta, \phi, t]\big) -$

$2\, i\, r\, \big(\psi 2^{(0,0,0,1)}[r, \theta, \phi, t] + \psi 3^{(1,0,0,0)}[r, \theta, \phi, t]\big) -$

$2\, i\, \big(\psi 3[r, \theta, \phi, t] - 2\, M\, \psi 3^{(1,0,0,0)}[r, \theta, \phi, t]\big) \Big) \Big),$

$\dfrac{1}{2\sqrt{1-\frac{2M}{r}}\, r^2} \Big(\Big(3\, i\, M - 2\, i\, r - \sqrt{1-\dfrac{2M}{r}}\, r\, \mathrm{Cot}[\theta]\Big) \psi 2[r, \theta, \phi, t] -$

$2\, r\Big(m\, \sqrt{1-\dfrac{2M}{r}}\, r\, \psi 3[r, \theta, \phi, t] + i\, r\, \psi 3^{(0,0,0,1)}[r, \theta, \phi, t] + \sqrt{1-\dfrac{2M}{r}}$

$\big(i\, \mathrm{Csc}[\theta]\, \psi 1^{(0,0,1,0)}[r, \theta, \phi, t] + \psi 2^{(0,1,0,0)}[r, \theta, \phi, t]\big) - i\, (2\, M - r)\, \psi 2^{(1,0,0,0)}[r, \theta, \phi, t] \Big) \Big),$

$\dfrac{1}{2\sqrt{1-\frac{2M}{r}}\, r^2} \Big(\Big(3\, i\, M - 2\, i\, r + \sqrt{1-\dfrac{2M}{r}}\, r\, \mathrm{Cot}[\theta]\Big) \psi 1[r, \theta, \phi, t] +$

$2\, r\Big(-m\, \sqrt{1-\dfrac{2M}{r}}\, r\, \psi 4[r, \theta, \phi, t] - i\, r\, \psi 4^{(0,0,0,1)}[r, \theta, \phi, t] + \sqrt{1-\dfrac{2M}{r}}$

$\big(i\, \mathrm{Csc}[\theta]\, \psi 2^{(0,0,1,0)}[r, \theta, \phi, t] + \psi 1^{(0,1,0,0)}[r, \theta, \phi, t]\big) + i\, (2\, M - r)\, \psi 1^{(1,0,0,0)}[r, \theta, \phi, t] \Big) \Big) \Big\}$

Fig. E.9 This is page 8 of the CARTAN evaluation of the F-I coefficients. Here we have the Dirac equations separated by commas

E.2   Using CARTAN in the N-P Approach

In this subsection we show how to use CARTAN to obtain the results of Sect. 4.2 mainly the NP spin coefficients. We reproduce in the figures below the relevant pages from the CARTAN files and comment on them.

```
(* $Id: Schwarzschild_NP.np.v 1.1 2002/02/18 12:09:03 soleng Exp $ *)

(* *********************************************************** *)
(*                                                             *)
(*              CARTAN library of geometries                   *)
(*                   (c) H. H. Soleng                          *)
(*                                                             *)
(*              Schwarzschild geometry                         *)
(*              ----------------------                         *)
(*                                                             *)
(*    We use the N-P formalism and compare with                *)
(*    Chandrasekhar's calculations pp. 134-135.                *)
(* *********************************************************** *)


(*    Newman-Penrose tetrad frame                              *)


nDim = 4

Coordinates = {r, th, ph, t}

OrientationOfFrame = 1

X[r]=1-(2*M)/r


Tetrad = {{1/Sqrt[2], 0, 0, X[r]/Sqrt[2]},
    {-1/(Sqrt[2]*X[r]), 0, 0, 1/Sqrt[2]},
    {0, r/Sqrt[2], -(I*r*Sin[th])/Sqrt[2], 0},
    {0, r/Sqrt[2], (I*r*Sin[th])/Sqrt[2], 0}}

FlatMetric = {{{0, 1, 0, 0}, {1, 0, 0, 0}, {0, 0, 0, -1}, {0, 0, -1, 0}},
    {1, 1}}

Riemannian = True

Torsion = "The manifold is Riemannian."
```

Fig. E.10 This is library file of the CARTAN evaluation of the N-P coefficients. The tetrad input is calculated using Eq. (3.91), moreover $\lambda^A = \left(\lambda^A{}_r, \lambda^A{}_\theta, \lambda^A{}_\phi, \lambda^A{}_t \right)$

```
Input File : Schwarzschild_NP.np
Note that the metric signature is (-2).  In this file we duplicate Chandrasekhar's calculations pp. 134-135, using a somewhat
modified null tetrad.
```

In[ ]:=
```
th = θ;
ph = φ;
```

In[ ]:=
```
Unprotect[RiemannR]
Remove[RiemannR]
Unprotect[Symmetrize]
Remove[Symmetrize]
```

Out[ ]=
```
{RiemannR}
```

Out[ ]=
```
{Symmetrize}
```

In[ ]:=
```
<< cartan.m
```

```
----------------------------------------

    Tensors in Physics: CARTAN 1.8
    by Harald H. Soleng, December 28, 2011
    Copyright © 2002-2011 by Ad Infinitum
    http://www.adinfinitum.no/cartan
----------------------------------------
```

```
Geometry given by:   1) interactive input
```

```
                     2) input from file
```

```
Selection [1]: 2
```

```
Show the structure coefficients [N]? N
```

```
Show the connection [N]? N
```

```
Show the metric [N]? Y
```

Fig. E.11 This is page 1 of the CARTAN evaluation of the N-P coefficients. Using the library file of Fig. E.10 we had to define at the top the variables th $= \theta$, ph $= \phi$ in order for the results look better. We did not ask CARTAN to evaluate the structure coefficients or the connection

$$g_{rr} = \frac{r}{2M - r}$$

$$g_{\theta\theta} = -r^2$$

$$g_{\phi\phi} = -r^2 \sin[\theta]^2$$

$$g_{tt} = -\frac{2M - r}{r}$$

Fig. E.12 This is page 2 of the CARTAN evaluation of the N-P coefficients. We have obtained the metric

The coefficients below differ from Chandrasekhar's results on p. 135 only by some $\sqrt{2}$.

In[ ]:= SSpinCoeff

Non-vanishing Newman-Penrose spin coefficients:

$$\alpha = -\frac{\cot[\theta]}{2\sqrt{2}\, r}$$

$$\beta = \frac{\cot[\theta]}{2\sqrt{2}\, r}$$

$$\gamma = \frac{M}{\sqrt{2}\, r^2}$$

$$\mu = -\frac{-2M + r}{\sqrt{2}\, r^2}$$

$$\rho = -\frac{1}{\sqrt{2}\, r}$$

Fig. E.13 This is page 3 of the CARTAN evaluation of the N-P coefficients. We have now obtained the Newman-Penrose spin coefficients of Eqs. (4.23)–(4.27). In order to obtain the four Dirac equations, (4.28)–(4.31), we have to use Eqs. (3.111) and (3.113), CARTAN will not do this part

Appendix F
Solutions to the Exercises

Exercise 1

$$\omega_{AB\mu} = e_{A\beta} \nabla_\mu e_B{}^\beta , \tag{F.1}$$

$$\omega_{BA\mu} = -e_{A\beta} \nabla_\mu e_B{}^\beta . \tag{F.2}$$

We relabel $A \leftrightarrow B$,

$$\omega_{AB\mu} = -e_{B\beta} \nabla_\mu e_A{}^\beta , \tag{F.3}$$

where

$$e_A{}^\beta = \eta_{CA} g^{\alpha\beta} e^C{}_\alpha , \tag{F.4}$$

thus

$$\omega_{AB\mu} = -e_{B\beta} \eta_{CA} g^{\alpha\beta} \nabla_\mu e^C{}_\alpha , \tag{F.5}$$
$$= -e_B{}^\alpha \eta_{CA} \nabla_\mu e^C{}_\alpha . \tag{F.6}$$

Relabel $A \leftrightarrow D$, then

$$\omega_{DB\mu} = -e_B{}^\alpha \eta_{CD} \nabla_\mu e^C{}_\alpha . \tag{F.7}$$
$$\omega^A{}_{B\mu} = \eta^{AD} \eta_{CD} \left(-e_B{}^\alpha \nabla_\mu e^C{}_\alpha \right) , \tag{F.8}$$
$$= \delta^A{}_C \left(-e_B{}^\alpha \nabla_\mu e^C{}_\alpha \right) , \tag{F.9}$$
$$= -e_B{}^\alpha \nabla_\mu e^A{}_\alpha , \tag{F.10}$$

which is Eq. (3.23).

Exercise 2 Starting with Eq. (3.31), we have

$$\left[i\bar{\gamma}^{\mu}(I\partial_{\mu} + \Gamma_{\mu}) - mI\right]\psi = 0, \tag{F.11}$$

$$\left[i\bar{\gamma}^{t}(I\partial_{t} + \Gamma_{t}) + i\bar{\gamma}^{k}(I\partial_{k} + \Gamma_{k}) - mI\right]\psi = 0, \tag{F.12}$$

where $k = (x^1, x^2, x^3)$. Now we multiply on the left by $\bar{\gamma}^{t}$. We choose our $\epsilon = +1$ in Eq. (3.5), thus

$$\left(\bar{\gamma}^{t}\right)^{2} = g^{tt} I. \tag{F.13}$$

Thus we obtain,

$$\left[ig^{tt}I(I\partial_{t} + \Gamma_{t}) + i\bar{\gamma}^{t}\bar{\gamma}^{k}(I\partial_{k} + \Gamma_{k}) - \bar{\gamma}^{t}m\right]\psi = 0. \tag{F.14}$$

We now divide by g^{tt} and obtain

$$i\partial_{t}\psi = \left[-i\Gamma_{t} - \frac{i}{g^{tt}}\bar{\gamma}^{t}\bar{\gamma}^{k}(I\partial_{k} + \Gamma_{k}) + \frac{m}{g^{tt}}\bar{\gamma}^{t}\right]\psi. \tag{F.15}$$

Therefore

$$H = -\frac{i}{g^{tt}}\bar{\gamma}^{t}\bar{\gamma}^{k}(I\partial_{k} + \Gamma_{k}) - i\Gamma_{t} + \frac{m}{g^{tt}}\bar{\gamma}^{t}, \tag{F.16}$$

where $k = (x^1, x^2, x^2)$.

Exercise 3 We shall calculate on of the matrix elements. From Eq. (3.87) we have that

$$\zeta_{12} = \lambda_1{}^{\alpha} \lambda_2{}^{\beta} g_{\alpha\beta} \tag{F.17}$$

$$= \frac{1}{\sqrt{2}}(e_0{}^{\alpha} + e_3{}^{\alpha})\frac{1}{\sqrt{2}}(e_0{}^{\beta} - e_3{}^{\beta}) g_{\alpha\beta}, \tag{F.18}$$

$$= \frac{1}{2}\left(e_0{}^{\alpha}e_0{}^{\beta}g_{\alpha\beta} - e_3{}^{\alpha}e_3{}^{\beta}g_{\alpha\beta}\right), \tag{F.19}$$

$$= \frac{1}{2}(1+1) = 1. \tag{F.20}$$

Exercise 4 From Eq. (3.89) we have

$$g^{\alpha\beta} = \zeta^{AB} \lambda_A{}^{\alpha} \lambda_B{}^{\beta}, \tag{F.21}$$

$$= \zeta^{12}\lambda_1{}^{\alpha}\lambda_2{}^{\beta} + \zeta^{21}\lambda_2{}^{\alpha}\lambda_1{}^{\beta} + \zeta^{34}\lambda_3{}^{\alpha}\lambda_4{}^{\beta} + \zeta^{43}\lambda_4{}^{\alpha}\lambda_3{}^{\beta}, \tag{F.22}$$

$$= \lambda_1{}^{\alpha}\lambda_2{}^{\beta} + \lambda_2{}^{\alpha}\lambda_1{}^{\beta} - \lambda_3{}^{\alpha}\lambda_4{}^{\beta} - \lambda_4{}^{\alpha}\lambda_3{}^{\beta}. \tag{F.23}$$

$$= l^{\alpha}n^{\beta} + n^{\alpha}l^{\beta} - m^{\alpha}\bar{m}^{\beta} - \bar{m}^{\alpha}m^{\beta}. \tag{F.24}$$

Exercise 5 Multiply Eq. (A.3) on both sides by $e_C{}^\alpha e_D{}^\beta$, then

$$e_C{}^\alpha e_D{}^\beta g_{\alpha\beta} = e_C{}^\alpha e_D{}^\beta e^A{}_\alpha e^B{}_\beta \eta_{AB},$$ (F.25)

$$= \delta^A_C \delta^B_D \eta_{AB} = \eta_{CD}.$$ (F.26)

Exercise 6 (a) We can use Eq. (3.29) and (3.30) rather than starting with Eq. (3.3), in order to write the Dirac equation since we know that the $\Gamma_C = 0$, thus we have

$$i\gamma^C h_C \Phi - m\Phi = 0,$$ (F.27)

where Φ is here is any one of the spinors of Eq. (D.14), the γ^C are given by Eq. (D.18), and the h_C are given by Eqs. (A.17)–(A.20).

(b) We know that each component of the spinor Φ transforms as a scalar function under general coordinate transformations, so now we transform the spinor to spherical coordinates. We note that the coordinates t, x, y, z, appear explicitly only in the exponential of Eq. (F.28).

$$\Phi^{(+)(1)}(t, x, y, z) = \frac{N}{\sqrt{2}} \begin{pmatrix} 1 - \dfrac{p_z}{p_t + m} \\[2mm] -\dfrac{p_x + ip_y}{p_t + m} \\[2mm] -1 - \dfrac{p_z}{p_t + m} \\[2mm] -\dfrac{p_x + ip_y}{p_t + m} \end{pmatrix} e^{-ip_\mu x^\mu},$$ (F.28)

We do not transform anything else. The exponent is transformed to

$$- i\left(p_t t + p_x r \sin\theta \cos\phi + p_y r \sin\theta \sin\phi + p_z \cos\theta\right).$$ (F.29)

Now we substitute the transformed $\Phi^{(+)(1)}(t, r, \theta, \phi)$ into Eq. (F.27). The details are a little tedious. We write below the first equation

$$\frac{1}{r}\left[e^{-i\phi}\left(\frac{1}{\sin\theta}\partial_\phi + i\cos\theta\,\partial_\theta + ir\sin\theta\,\partial_r\right)\Phi_4 \right.$$
$$\left. -i\left(r\,\partial_t + \sin\theta\,\partial_\theta - r\cos\theta\,\partial_r\right)\Phi_3\right] - m\Phi_1 = 0, \quad \text{(F.30)}$$

where

$$\Phi^{(+)(1)}(t, r, \theta, \phi) = \begin{pmatrix} \Phi_1 \\ \Phi_2 \\ \Phi_3 \\ \Phi_4 \end{pmatrix}. \tag{F.31}$$

We find that it vanishes as expected.

Exercise 7 We substitute Eq. (D.3) and $\phi = U\psi$ into equation

$$(\gamma'^{\mu} p_{\mu} - mI)\phi = 0, \tag{F.32}$$

then Eq. (F.32) becomes

$$\left(U\gamma^{\mu}U^{-1} p_{\mu} - mI\right) U\psi = 0, \tag{F.33}$$

$$U\left(\gamma^{\mu} p_{\mu} - mI\right)\psi = 0. \tag{F.34}$$

Exercise 8 In the solution below, $q_0 = q_t$, $p_0 = p_t$, and so forth. From the last line of Eq. (D.52) we have,

$$\bar{v}^{(\beta)}(q)\left(-\not{q}\gamma^0 + \gamma^0\not{p}\right) u^{(\alpha)}(p), \tag{F.35}$$

$$= \bar{v}^{(\beta)}(q)\left(-q_{\mu}\gamma^{\mu}\gamma^0 + \gamma^0 p_{\mu}\gamma^{\mu}\right) u^{(\alpha)}(p), \tag{F.36}$$

$$= \bar{v}^{(\beta)}(q)\left(-\left(q_0\gamma^0 + q_i\gamma^i\right)\gamma^0 + \gamma^0\left(p_0\gamma^0 + p_i\gamma^i\right)\right) u^{(\alpha)}(p), \tag{F.37}$$

$$= \bar{v}^{(\beta)}(q)\left(-\left(q^0\gamma^0 - q^i\gamma^i\right)\gamma^0 + \gamma^0\left(p^0\gamma^0 - p^i\gamma^i\right)\right) u^{(\alpha)}(p), \tag{F.38}$$

$$= \bar{v}^{(\beta)}(q)\left(-\left(p^0\gamma^0 + p^i\gamma^i\right)\gamma^0 + \gamma^0\left(p^0\gamma^0 - p^i\gamma^i\right)\right) u^{(\alpha)}(p), \tag{F.39}$$

$$= \bar{v}^{(\beta)}(q)\left(-p^i\gamma^i\gamma^0 - p^i\gamma^0\gamma^i\right) u^{(\alpha)}(p) = 0. \tag{F.40}$$

Exercise 9 In the solution below, $q_0 = q_t$, $p_0 = p_t$, and so forth. From Eq. (D.55) we have, using Eq. (D.53), that

$$\left(\Psi_{p'}^{(\alpha)} | \Psi_p^{(\beta)}\right) = \mathbb{N}^2 \int \bar{u}^{(\alpha)}\left(p'\right) \gamma^0 u^{(\beta)}(p) e^{i\left(p'_{\mu} - p_{\mu}\right)x^{\mu}} d^3x, \tag{F.41}$$

$$= \mathbb{N}^2 \int \bar{u}^{(\alpha)}\left(p'\right) \gamma^0 u^{(\beta)}(p) e^{i\left(p'_0 - p_0\right)x^0 + i\left(p - p'\right)x} d^3x. \tag{F.42}$$

Now we use the result in Eq. (D.46) and obtain

$$\left(\Psi_{p'}^{(\alpha)} | \Psi_{p}^{(\beta)} \right) = \mathbb{N}^2 \frac{p^t}{m} \delta^{\alpha\beta} e^{i(p_0' - p_0)x^0} \int e^{i(p - p')x} d^3x \,, \tag{F.43}$$

$$= \mathbb{N}^2 \frac{p^t}{m} \delta^{\alpha\beta} e^{i(p_0' - p_0)x^0} (2\pi)^3 \delta^3 \left(p - p' \right) \,. \tag{F.44}$$

Since $p_t = p^0 = +\sqrt{p^2 + m^2}$, we see that the $\delta^3 \left(p - p' \right)$ in Eq. (F.44), guarantees that the energies of the two spinors involved, are also equal. So we can get rid of the exponential in Eq. (F.44).

References

1. S. Carroll, *Spacetime and Geometry, an Introduction to General Relativity* (Addison-Wesley, San Francisco, 2004)
2. M. Christodoulou, A. Riello, C. Rovelli, How to detect anti-spacetime. Int. J. Mod. Phys. D **21**, 1242014 (2012)
3. S. Weinberg, *Gravitation and Cosmology: Principles and Applications of the General Theory of Relativity* (Wiley, New York, 1972)
4. D. Klein, P. Collas, General transformation formulas for Fermi-Walker coordinates. Class Quantum Grav. **25**, 145019 (2008), arXiv:0712.3838
5. C. Chicone, B. Mashhoon, Explicit Fermi coordinates and tidal dynamics in de Sitter and Gödel spacetimes. Phys. Rev. D **74**, 064019 (2006)
6. D. Klein, P. Collas, Exact Fermi coordinates for a class of space-times. J. Math. Phys. **51**, 022501 (2010)
7. D. Bini, A. Geralico, R.T. Jantzen, Fermi coordinates in Schwarzschild spacetime: closed form expressions. Gen. Relativ. Gravit. **43**, 1837–1853 (2011)
8. D. Klein, Maximal Fermi charts and geometry of inflationary universes. Ann Henri Poincaré (2012). https://doi.org/10.1007/s00023-012-0227-3
9. D.R. Brill, J.A. Wheeler, Interaction of neutrinos and gravitational fields. Rev. Mod. Phys. **29**, 465–479 (1957)
10. V.M. Villalba, U. Percoco, Separation of variables and exact solution to Dirac and Weyl equations in Robertson-Walker space-times. J. Math Phys. **31**, 715–720 (1990)
11. G.V. Shishkin, Some exact solutions of the Dirac equation in gravitational fields. Class. Quantum Grav. **8**, 175–185 (1991)
12. P.B. Groves, P.R. Anderson, E.D. Carlson, Method to compute the stress-energy tensor for the massless spin $\frac{1}{2}$ field in a general static spherically symmetric spacetime. Phys. Rev. D **66**, 124017 (2002)
13. G. de A. Marques, V.B Bezerra, S.G. Fernandes, Exact solution of the Dirac equation for a Coulomb and scalar potentials in the gravitational field of a cosmic string. Phys. Lett. A **341**, 39-47 (2005)
14. X.-N. Zhou, X.-L. Du, K. Yang, Y.-X. Liu, Dirac dynamical resonance states around Schwarzschild black holes. Phys. Rev. D **89**, 043006 (2014)
15. I.I. Cotaescu, C. Crucean, C.A. Sporea, Partial wave analysis of the Dirac fermions scattered from Schwarzschild black holes. Eur. Phys. J. C **76:102**, 1–19 (2016) Partial wave analysis of the Dirac., https://link.springer.com/article/10.1140/epjc/s10052-016-3936-9
16. E.A. Lord, *Tensors, Relativity and Cosmology* (Tata McGraw-Hill Publishing Co., Ltd., New Delhi, 1976)
17. R.M. Wald, *General Relativity* (U. of Chicago Press, Chicago, 1984), pp. 351–360
18. L.H. Ryder, *Quantum Field Theory*, 2nd edn. (Cambridge U. Press, New York, 1996), pp. 436–7

19. G.L. Naber, *The Geometry of Minkowski Spacetime* (Springer-Verlag, New York, 1992)
20. L. Parker, D. Toms *Quantum Field Theory in Curved Spacetime* (Cambridge U. Press, Cambridge, 2009)
21. H. Yamamoto, Quantum Field Theory for Non-specialists. Unpublished Lecture notes, Ch. 1–6 (2012), http://epx.phys.tohoku.ac.jp/~yhitoshi/particleweb/particle.html
22. P. Collas, D. Klein, Dirac particles in a gravitational shock wave. Class Quantum Grav. **35**, 125006 (2018). https://doi.org/10.1088/1361-6382/aac144
23. L. Parker, One-electron atom as a probe of spacetime curvature. Phys. Rev. D **22**, 1922–1934 (1980)
24. J.M. Jauch, F. Rohrlich, *The Theory of Photons and Electrons*, 2nd edn. (Springer-Verlag, New York, 1976)
25. S.P. Gavrilov, D.M. Gitman, J.L. Tomazelli, Comments on spin operators and spin-polarization states of 2 + 1 fermions. Eur. Phys. J. **39**, 245–248 (2005)
26. A. Peres, Spinor fields in generally covariant theories. Nuovo Cimento **24**, 389–452 (1962)
27. L. Ryder, *Introduction to General Relativity* (Cambridge U. Press, Cambridge, 2009)
28. M.N. Hounkonnou, J.E.B. Mendy, Exact solutions of the Dirac equation in a nonfactorizable metric. J. Math. Phys. **40**, 3827–3842 (1999)
29. D.S. Shirokov, Pauli's theorem in the description of n-dimensional spinors in the Clifford algebra formalism. Theor. Math. Phys. **175**, 454–474 (2013)
30. D.S. Shirokov, Clifford algebras and their applications to Lie groups and spinors, pp. 1–44. 20 Jan 2018. arXiv:1709.06608v2
31. W. Greiner, *Relativistic Quantum Mechanics, Wave Equations*, 3rd edn. (Springer-Verlag, New York, 2000)
32. T. Ohlsson, *Relativistic Quantum Mechanics* (Cambridge U. Press, Cambridge, 2011)
33. A.M. Steane, An introduction to spinors, pp. 1–23. 13 Dec 2013. arXiv:1312.3824v1
34. M.E. Peskin, D.V. Schroeder, *An Introduction to Quantum Field Theory*, (Perseus Books, Reading, 1995)
35. C. Itzykson, J.-B. Zuber, *Quantum Field Theory* (McGraw-Hill Inc., New York, 1980)
36. H.H. Soleng, *Tensors in Physics: the Mathematica package Cartan version 1.8* (Ad Infinitum AS, Fetsund, Norway, 2011). The Mathematica package is available at https://store.wolfram.com/view/book/D0709.str

Index

© The Author(s), under exclusive license to Springer Nature Switzerland AG 2019   107
P. Collas and D. Klein, *The Dirac Equation in Curved Spacetime*,
SpringerBriefs in Physics, https://doi.org/10.1007/978-3-030-14825-6

Printed in the United States
By Bookmasters